Lecture Notes in Mathematics

Edited by A. Dold and B. Eckmann

443

Michel Lazard

Commutative Formal Groups

Springer-Verlag
Berlin · Heidelberg · New York 1975

Prof. Dr. Michel Lazard
2 rue Boutarel
F-75004 Paris

AMS Subject Classifications (1970): 12B25, 14L05, 20J05, 20J15, 20K40, 22E05

ISBN 3-540-07145-8 Springer-Verlag Berlin · Heidelberg · New York
ISBN 0-387-07145-8 Springer-Verlag New York · Heidelberg · Berlin

Offsetdruck: Julius Beltz, Hemsbach/Bergstr.

CONTENTS

FOREWORD 4

GENERAL CONVENTIONS 6

CHAPTER I

FORMAL VARIETIES 8

1. The categories $\underline{\underline{nil}}(K)$ and $\underline{\underline{nil}}(K,n)$ 8
2. Functors in pointed sets 9
3. Models and their morphisms 10
4. Formal varieties 13
5. Formal modules 14
6. The functor $\mathfrak{T}$ (tangent) 17
7. The composition lemma 19
8. The isomorphism theorem 22
9. The fibration $J_q \to J_{q-1}$ 23
10. The functor $\mathfrak{C}$ (curves) 25
11. Changes of rings 28

CHAPTER II

FORMAL GROUPS AND BUDS 31

1. Groups in categories 31
2. Group laws and formal groups 33
3. The $\underline{\underline{Q}}$ theorem, the lift theorem, and the general existence of ghosts 37
4. Buds and extensions 39
5. Obstructions 40
6. The 1-dimensional symmetric 2-cocycle lemma 44
7. Curvilinear group laws and buds 47
8. End of the proofs of the theorems 51
9. A digression concerning non-commutative groups 54

CHAPTER III

THE GENERAL EQUIVALENCE OF CATEGORIES 57

1. Definition of W^+ and $\hat{W}^+$ from $\mathfrak{C}(\underline{\underline{G}}_m)$ 57
2. The formal group homomorphism $\underline{\underline{w}} : \hat{W}^+ \to D_+^{(\underline{\underline{p}})}$ 61
3. The F_n operators 64

4. The representation theorem: $\mathfrak{C}(G) \simeq \mathrm{Hom}(\hat{W}^+, G)$ 68

5. Introducing the ring of operators Cart(K) 70

6. Curves in formal groups and representations of operators 72

7. Continuity, uniform modules, reduced modules 75

8. How $\mathfrak{C}$ is fully faithful on formal groups 79

9. Some corollaries of the representation theorem 81

10. The existence theorem: preliminaries with a uniform module 83

11. The existence theorem: end of the proof 86

CHAPTER IV

THE SPECIAL EQUIVALENCES OF CATEGORIES 92

1. The functor Cart and the commuting relations 92

2. An axiomatic description of the functor Cart_S 94

3. Properties and existence of Cart_S are derived from the matrix representation κ' 96

4. The ring $W_S(K)$ and its integers 101

5. Uniform and reduced $\mathrm{Cart}_S(K)$-modules 105

6. The functorial ring homomorphism $\kappa_{T,S}$ 107

7. The category of S-typical groups 110

8. The reduction theorem 114

9. An example: between Artin-Hasse and multiplication 116

CHAPTER V

THE STRUCTURE THEOREM AND ITS CONSEQUENCES 118

1. Free uniform $\mathrm{Cart}_S(K)$-modules and types. 118

2. Slopes and types 121

3. The structure theorem 124

4. A second proof of the existence theorem 129

5. Presentations, structural constants, morphisms. 133

6. Tensor products 136

7. Definition and general properties of reduced derivatives 141

8. S-typical groups over S-torsion-free rings 144

9. Some examples 149

10. The parametrization of curvilinear group laws 152

11. A digression concerning derivatives 159

CHAPTER VI

ON FORMAL GROUPS IN CHARACTERISTIC p 163

1. Notations for the local case 163
2. The special features of characteristic p 167
3. Fields and perfect fields 170
4. Finite dimension and isogeny 174
5. Unipotent formal groups 179
6. Spectral decomposition of semi-linear automorphisms of finite $\tilde{E}$-modules 183
7. Formal groups of finite height 189
8. Isoclinal formal groups over an algebraically closed field 196

CHAPTER VII

EXTENDING AND LIFTING SOME FORMAL GROUPS 199

1. Extensions with additive kernels of formal groups of finite height 199
2. The universal extension with additive kernel in characteristic p 204
3. The reduction modulo p of p-typical groups 209
4. On some ring homomorphisms $A \to W(A)$ 212
5. A p-adic lemma 215
6. Reduction from W(k) to k (a perfect field of characteristic p) 218
7. Lifts from k over W(k) 223
8. The universal extension with additive kernel in characteristic 0 228

QUOTED REFERENCES 231

INDEX 232

Foreword

The prerequisites of the present book are limited to a good acquaintance with "graduate" algebra and with the elements of "ultrametric analysis", such as filtrations, valuations, completions, etc.

The classical analogue of a "formal variety" is a germ of analytic manifold, not a whole manifold. The theory of formal groups corresponds to the local theory of Lie groups. Only commutative formal groups are studied here.

The book is by no means a complete survey of the present state of the theory: theorems of the greatest interest, such as Lubin-Tate's [13], have not been included.

Chapters I to V give a systematic development of my notes "Sur les théorèmes fondamentaux des groupes formels commutatifs" [9], i.e. a covariant construction based on Cartier's Frobenius operators (see III.3) together with the general lift theorem (see II.3). Whether covariance does or does not differ from contracontravariance is left to the Reader's jugement.

Chapters I to III form a whole, leading from the basic definitions to a very general theorem: the equivalence of the category of commutative formal groups over a ring K with that of a class of topological modules over a big topological ring, denoted by Cart(K). The study of this ring Cart(K) is not pushed further than needed to prove the theorem.

Chapter IV explains how, in some cases, the ring Cart(K) may be replaced by some "smaller" ring $\mathrm{Cart}_S(K)$, where S denotes a set of primes.

Chapter V justifies the equivalences of categories of formal groups with modules, by showing that the modules are actually more manageable than the formal groups.

Chapter VI contains Dieudonné's results [7], and more.

Chapter VII could have been called "Cartier's 1972 seminar revisited" or "the trick of the (W,F)-linear section". For an introduction to that chapter, the Reader may consult [5].

Historical considerations are reduced to this: it was J.-P. Serre who prompted me to publish Cartier's results.

Apologies are offered to The English Language, for mixing unwittingly not only cats and bats, but also this and that, and worse. Nevertheless, I hope that some meaning has been preserved.

It is a pleasure to thank the following people:

Karl W. GRUENBERG from Queen Mary College, London, who was the friendly instigator of the present paper;

Beno ECKMANN, Professor at the E.T.H., Zürich, and Editor of the Springer Notes, who, in each capacity, helped to give shape to the project;

Mrs. M. von WILDEMANN who made everything easy for me at the Zürich Institute;

finally Miss R. BOLLER, who typed the manuscript so splendidly.

M. Lazard

(manuscript dated: December 1973 - April 1974).

GENERAL CONVENTIONS

1. The <u>constants</u>, such as $\underline{\underline{N}}$, $\underline{\underline{Z}}$, $\underline{\underline{Q}}$, etc. are as in N. Bourbaki's treatise, but for one exception. Namely the set of the rational integers ≥ 1 is denoted by $\underline{\underline{P}}$, not by $\underline{\underline{N}}^*$. According to a convention to be introduced in (IV. 2.1), P denotes the set of all rational primes.

As usual "iff" means "if and only if".

2. "<u>Almost all</u>" elements of a set X verify some condition P iff the set of the $x \in X$, such that $P(x)$ does not hold, is finite.

In a topological space X, an <u>indexed set</u> $(x_i)_{i \in I}$ <u>converges towards a limit</u> $x \in X$ iff, for any neighbourhood N of x in X, $x_i \in N$ for almost all $i \in I$.

3. A <u>pointed set</u> is a set E together with a given point in E, usually denoted by 0_E or even by 0. The <u>category of pointed sets</u>, denoted by "$\underline{\underline{pts}}$" is defined by taking as morphisms the maps $f : E \to F$ such that $f(0_E) = 0_F$.

4. The product of an indexed set of pointed sets, $E = \Pi_{i \in I} E_i$ is a pointed set (the natural maps $E \to E_i$ being morphisms). When $E_i = A$ for every $i \in I$, then the product is written as a power: A^I.

The <u>support</u> of an element $x = (x_i)_{i \in I} \in E$ in a product is the set of indices $i \in I$ such that $x_i \neq 0$. The support of x is denoted by $\mathrm{supp}(x)$.

The <u>restricted product</u> of an indexed set of pointed sets, denoted by $\Pi^*_{i \in I} E_i$, is the subset of elements with <u>finite support</u> in the (unrestricted) product $\Pi_{i \in I} E_i$. When $E_i = A$ for any $i \in I$, the restricted product is a "<u>restricted power</u>", <u>written</u>

$A^{(I)}$. Of course, the difference is significant only when I is infinite.

5. Monomials are assumed to be associative and commutative. We shall often consider monomials in a (possibly) infinite set of variables, say $(x_i)_{i \in I}$. Then the monomials will always remain finite products, i.e. almost all exponents will be 0. We write x^α for $\Pi_{i \in I} x_i^{\alpha_i}$, where $\alpha = (\alpha_i)_{i \in I} \in \underline{\underline{N}}^{(I)}$. We say that α is the "multi-degree" of x^α ; the total degree is $|\alpha| = \Sigma_{i \in I} \alpha_i \in \underline{\underline{N}}$. We put $\alpha! = \Pi_{i \in I} \alpha_i!$

6. When every variable x_i has been assigned some element w_i of an additive monoïd W, then the weight of a monomial x^α is defined as $\Sigma_{i \in I} \alpha_i w_i \in W$. Total or partial degrees are special cases of weights. An isobaric polynomial is a linear combination of monomials with given weight. Products of isobaric polynomials are again isobaric.

7. Isobaric polynomials of weight 0 form a subring (of the ring of all polynomials). This remark is useful when considering polynomials with indeterminate coefficients, such as $P(x) = \Sigma_\alpha c_\alpha x^\alpha$; if the coefficient c_α receives the weight $-\alpha$, then $P(x)$ becomes an isobaric polynomial of weight 0. If $Q(x) = \Sigma_\alpha d_\alpha x^\alpha$ is a polynomial in $P(x)$, then d_α is an isobaric polynomial of weight α in the coefficients c_β of respective weights β. As an example, let us consider the identity of formal series

$$1 - \theta \Sigma_{n \in \underline{\underline{N}}} a_n t^n = \Pi_{m,n \in \underline{\underline{N}}} (1 - b_{m,n} t^m \theta^n) .$$

Once it is known that any choice of the coefficients a_n, in any commutative ring, defines the coefficients $b_{m,n}$, then the remark implies that every $b_{m,n}$ is a polynomial of weight $(m,n) \in \underline{\underline{N}}^2$ in the a_n, when a_i receives the weight $(i,1)$.

CHAPTER I

FORMAL VARIETIES

1. *The categories* $\mathbf{nil}(K)$ *and* $\mathbf{nil}(K,n)$

1.1 By a *basic ring* K, we shall mean an associative and commutative ring, with unit. Changes of basic rings will be considered later.

1.2 Together with K, we introduce the category $\mathbf{nil}(K)$ of associative and commutative *nilalgebras* over K (any x in any $A \in \mathbf{nil}(K)$ is nilpotent). Morphisms in $\mathbf{nil}(K)$ are K-algebras homomorphisms.

As there is no unit in a (nonzero) nilalgebra, we *define* A-*modules* as unitary $(K \oplus A)$-modules, where A is identified with the augmentation ideal in the supplemented K-algebra $K \oplus A$.

1.3 For each $n \in \mathbb{N}$, we denote by $\mathbf{nil}(K,n)$ the full subcategory of $\mathbf{nil}(K)$ defined by the property that every product of $n+1$ elements in any $A \in \mathbf{nil}(K,n)$ is 0. We have the strict inclusions

$$\mathbf{nil}(K,0) \subset \ldots \subset \mathbf{nil}(K,n) \subset \mathbf{nil}(K,n+1) \subset \ldots \subset \mathbf{nil}(K)$$

The category $\mathbf{nil}(K,0)$ contains only the zero algebra; $\mathbf{nil}(K,1)$ may be identified with the category of K-modules, products of two elements being always 0.

The union of the $\mathbf{nil}(K,n)$ is the category of *nilpotent* K-algebra, which we denote by $\mathbf{nilp}(K)$; note that $\mathbf{nilp}(K) \neq \mathbf{nil}(K)$.

1.4 *Any finitely generated subalgebra* B *of* $A \in \mathbf{nil}(K)$ *is nilpotent*.

For, if B is generated by $x_1,\dots,x_n \in A$, with

$x_i^{\alpha_i} = 0$ for $\alpha_i > n_i$, we put $n = \Sigma_i n_i$ and we have

$x^\alpha = 0$ for $|\alpha| > n$, so that $B \in \underline{\underline{nil}}(K,n)$.

1.5 The category $\underline{\underline{nil}}(K)$ does not contain free objects (because a free generator should have an arbitrarily large order of nilpotence, which is absurd), whereas each category $\underline{\underline{nil}}(K,n)$ does. More precisely, the set $(x_i)_{i\in I}$ generates freely in $\underline{\underline{nil}}(K,n)$ an algebra, where the monomials x^α are linearly independent for $0 < |\alpha| \leqslant n$ and 0 for $|\alpha| > n$.

2. Functors in pointed sets

2.1 A functor $V : \underline{\underline{nil}}(K) \to \underline{\underline{pts}}$ associates to any $A \in \underline{\underline{nil}}(K)$ a pointed set $V(A)$, and to any morphism $\varphi : A \to A'$ in $\underline{\underline{nil}}(K)$ a pointed map $V(\varphi) : V(A) \to V(A')$, with the conditions $V(Id) = Id$, $V(\varphi\circ\psi) = V(\varphi) \circ V(\psi)$.

2.2 A functorial morphism (or natural application), $f : V \to W$, of two such functors associates to any $A \in \underline{\underline{nil}}(K)$ a pointed map $f_A : V(A) \to W(A)$, with commutative diagrams: $f_{A'} \circ V(\varphi) = W(\varphi) \circ f_A$.

2.3 Given three functors $U,V,W : \underline{\underline{nil}}(K) \to \underline{\underline{pts}}$ and two functorial morphisms $f : U \to V$, $g : V \to W$, the composed functorial morphism

$g\circ f : U \to W$ is defined by $(g\circ f)_A = g_A \circ f_A$ for $A \in \underline{\underline{nil}}(K)$. Composition is associative.

2.4 Let $n \in \underline{\underline{N}}$. By restriction from $\underline{\underline{nil}}(K)$ to $\underline{\underline{nil}}(K,n)$, each $V : \underline{\underline{nil}}(K) \to \underline{\underline{pts}}$ gives a functor $J_n V : \underline{\underline{nil}}(K,n) \to \underline{\underline{pts}}$, and each $f : V \to W$ gives a functorial morphism

$J_n f : J_n V \to J_n W$. We call $J_n V$ and $J_n f$ the n-th <u>jets</u> of V and of f. We have $J_n(g \circ f) = J_n g \circ J_n f$.

2.5 We shall be interested only in special functors from $\underline{\underline{\text{nil}}}(K)$ to $\underline{\underline{\text{pts}}}$ for which the class $\mathfrak{M}(V,W)$ of functorial morphisms from V to W is a <u>set</u>. It will then be legitimate to consider the category, the object of which are those functors, and the morphisms of which are their functorial morphisms.

On a set $\mathfrak{M}(V,W)$ we shall then have <u>two uniform structures</u>, which we call just "topologies" (see III. 1.2).

2.6 The "<u>order</u>" topology is defined from the relations $J_n f = J_n f'$, with variable $n \in \underline{\underline{N}}$.

2.7 The "<u>simple</u>" topology is defined from the relations $f_A(x) = f'_A(x)$, with variables $A \in \underline{\underline{\text{nil}}}(K)$ and $x \in V(A)$.

2.8 Note that $\mathfrak{M}(V,W)$ is complete for the simple topology, and that a composition map $\mathfrak{M}(U,V) \times \mathfrak{M}(V,W) \to \mathfrak{M}(U,W)$ is continuous for the simple topology.

3. <u>Models and their morphisms</u>

3.1 Let I be a set. We define the <u>model indexed by</u> I as the functor $A \mapsto A^{(I)}$ from $\underline{\underline{\text{nil}}}(K)$ to $\underline{\underline{\text{pts}}}$, and we denote this functor by $D^{(I)}$.

In particular, D^n is the "n-space of points with nilpotent coordinates over K", and D is the forgetful functor $A(\in \underline{\underline{\text{nil}}}(K)) \mapsto A(\in \underline{\underline{\text{pts}}})$.

3.2 <u>The morphisms lemma</u>. <u>The functorial morphisms</u> $f : D^{(I)} \to D$ <u>are in one-to-one correspondence with the formal series</u>

$$f(x) = \Sigma_{\alpha \in \underline{\underline{N}}^{(I)},\, |\alpha|>0}\, c_\alpha x^\alpha \quad ,$$

<u>with coefficients</u> $c_\alpha \in K$ <u>subject to no condition</u>. (We shall often omit to write "$|\alpha| > 0$" by putting $c_o = 0$).

Proof.

3.3 For any $A \in \underline{\underline{\underline{nil}}}(K)$ and any $x = (x_i)_{i \in I} \in A^{(I)}$, almost all monomials x^α are 0 (see 1.4), so that a formal series actually defines a functorial morphism.

3.4 Conversely, let Λ be the set of pairs (J,n) where J is a <u>finite</u> subset of I and $n \in \underline{\underline{P}}$. For any $\lambda = (J,n) \in \Lambda$, let A_λ be freely generated in $\underline{\underline{\underline{nil}}}(K,n)$ by elements $x_{\lambda,i}$ for $i \in J$ (see 1.5). We define the point $x_\lambda = (x_{\lambda,i})_{i \in I}$ in $A_\lambda^{(I)}$ by putting $x_{\lambda,i} = 0$ for $i \notin J$.

Now, if $f : D^{(I)} \to D$ is a functorial morphism, we must have for each $\lambda = (J,n) \in \Lambda$,

3.5 $$f_{A_\lambda}(x_\lambda) = \Sigma_{\alpha \in \underline{\underline{N}}^{(I)}, 0<|\alpha| \leq n,\ \mathrm{supp}(\alpha) \subset J}\ c_{\lambda,\alpha}\, x_\lambda^\alpha ,$$

with uniquely determined $c_{\lambda,\alpha} \in K$. But, for any $A \in \underline{\underline{\underline{nil}}}(K)$ and any $x \in A^{(I)}$, there is a $\lambda \in \Lambda$ and a $\varphi \in \mathrm{Hom}(A_\lambda, A)$, such that $D^{(I)}(\varphi): A_\lambda^{(I)} \to A^{(I)}$ maps x_λ on x. The formula (3.5) and the functoriality of f imply

$$f_A(x) = \Sigma_{\alpha \in \underline{\underline{N}}^{(I)}, 0<|\alpha| \leq n,\ \mathrm{supp}(\alpha) \subset J}\ c_{\lambda,\alpha}\, x^\alpha .$$

By taking $x = x_{\lambda'}$, where $\lambda' = (J',n')$, $J' \subset J$, $n' \leq n$, we see that $c_{\lambda,\alpha} \in K$ is independent of $\lambda = (J,n)$, provided that $J \supset \mathrm{supp}(\alpha)$ and $n \geq |\alpha|$. Writing $c_{\lambda,\alpha} = c_\alpha$ for such λ, we have completed the proof.

3.6 For any two models, $D^{(I)}$, $D^{(J)}$, a functorial morphism $f : D^{(I)} \to D^{(J)}$ is defined by its components $f_j : D^{(I)} \to D$, $j \in J$. The <u>set</u> of such f will be denoted by $\mathfrak{M}(D^{(I)}, D^{(J)})$, and they will be called morphisms: we have just defined the <u>cate-</u>

gory of models, over K (see 2.5).

3.7 An important simplification takes place in the study of $D^{(I)}$ when I is finite. Namely we need not consider the points x_λ for all $\lambda = (J,n)$, but only for $\lambda = (I,n)$, with variable n; we can then write x_n instead of x_λ. In this case the order topology (2.6) coincides with the simple topology (2.7), on any set $\mathfrak{M}(D^{(I)},D^{(J)})$.

3.8 When I is infinite, those topologies do not coincide: the canonical projections $p_i: D^{(I)} \to D$ converge towards 0 for the simple topology, not for the order topology.

3.9 Without any assumption about I, the order topology is finer than the simple topology (that is: neighbourhoods are smaller in the first than in the second). This expresses only the fact that "points come from nilpotent points" in models (any $x \in A^{(I)}$ is a homomorphic image of some $x' \in A'^{(I)}$ with nilpotent A').

3.10 More precisely, a set $(f_j)_{j\in J}$ in $\mathfrak{M}(D^{(I)}, D)$, given by $f_j(x) = \Sigma_\alpha c_{j,\alpha} x^\alpha$ as in (3.2) converges towards zero: 1°) in the simple topology iff, for any $\alpha \in \underline{\underline{N}}^{(I)}$, $c_{j,\alpha} = 0$ for almost all $j \in J$; 2°) in the order topology iff, for any $n \in \underline{\underline{P}}$ and all $\alpha \in \underline{\underline{N}}^{(I)}$ such that $|\alpha| \leq n$, $c_{j,\alpha} = 0$ for almost all $j \in J$ (the number of α with $|\alpha| \leq n$ is finite iff I is finite).

3.11 The morphisms $f : D^{(I)} \to D^{(J)}$ are defined by their components $(f_j)_{j\in J}$, subject to the condition of converging towards 0 in the simple topology. They will be written in a more concise way later (5.2).

4. Formal varieties

4.1 Definition. A <u>formal variety</u> V over K is a functor $V : \underline{\underline{nil}}(K) \to \underline{\underline{pts}}$, which is isomorphic to some model $D^{(I)}$. In other words, there exists a (functorial) morphism $f : D^{(I)} \to V$ such that $f_A : A^{(I)} \to V(A)$ is one-to-one for any $A \in \underline{\underline{nil}}(K)$. Such an $f : D^{(I)} \to V$ will be called a <u>coordinate system on</u> V, indexed by I.

4.2 The <u>category of formal varieties</u> is defined by taking functorial morphisms as morphisms (see 2.5 and 3.6). The reason why formal varieties are introduced, instead of only models (of which they are just isomorphic copies) is to regain the classical point of view, where a given geometric object may be described by different systems of coordinates.

4.3 The category of formal varieties contains <u>finite products</u> and <u>infinite restricted products</u>. Namely, if $(W_i)_{i \in I}$ is an indexed set of formal varieties, $\prod^*_{i \in I} W_i = W$ is the functor $A \mapsto \prod^*_{i \in I} W_i(A)$; a morphism $f : V \to W$ is defined by its component $f_i : V \to W_i$, and, for each $A \in nil(K)$ and each $x \in V(A)$, $f_{i,A}(x) \in W_i(A)$ has to be 0 for almost all i.

4.4 For a morphism from a product, say $f : U \times V \to W$, we shall use freely the functional notation and introduce variables, writing f as $f(x,y)$. Strictly speaking, only the f_A are functions, but this "abus de langage" is very convenient for writing composed morphisms.

4.5 For instance, we say that a morphism $f : V \times V \to V$ is <u>associative</u> iff $f(f(x,y),z) = f(x,f(y,z))$, and <u>commutative</u> iff $f(x,y) = f(y,x)$.

5. Formal modules

5.1 Let L be a free K-module. We obtain from L a formal variety in the following way: for each $A \in \underline{\underline{nil}}(K)$ and each morphism $\varphi : A \to A'$, the set $A \otimes_K L$ is associated to A and the map $\varphi \otimes Id_L : A \otimes L \to A' \otimes L$ is associated to φ. The choice of a basis $(e_i)_{i \in I}$ in L allows us to write any $x \in A \otimes L$ uniquely as $\Sigma_{i \in I}\, y_i \otimes e_i$ with $y = (y_i)_{i \in I} \in A^{(I)}$. So, apart from the language, the model $D^{(I)}$ is obtained in this way from the free module $K^{(I)}$ with indexed basis.

5.2 Let $f = (f_j)_{j \in J}$ be a morphism of models, $f : D^{(I)} \to D^{(J)}$. We have, for each $j \in J$,

$$f_j(x) = \Sigma_{\alpha \in \underline{\underline{N}}^{(I)}}\; c_{j,\alpha}\, x^{\alpha}$$

with coefficients $c_{j,\alpha} \in K$ (see 3.2), and, for each α, $c_{j,\alpha} = 0$ for almost all j (see 3.10). So we can put

$$c_{\alpha} = (c_{j,\alpha})_{j \in J} \in K^{(J)}$$

and write f as

$$f(x) = \Sigma_{\alpha \in \underline{\underline{N}}^{(I)}}\; c_{\alpha} x^{\alpha}\ , \text{ with } c_{\alpha} \in K^{(J)},\ c_0 = 0\ .$$

This is quite meaningful if we identify $A^{(J)}$ with $A \otimes K^{(J)}$ and $c_{\alpha} x^{\alpha}$ with $x^{\alpha} \otimes c_{\alpha}$.

5.3 The formal modules stand between models and formal varieties as shown in the following table.

	models	formal modules	formal varieties
notations	$D^{(I)}$, I: any set	L^+; L: any free K-module	V: any functor $\underline{\underline{nil}}(K) \to \underline{\underline{pts}}$ isomorphic to a model
changes of coordinates	none	K-linear changes	any change over K
forgetting some structure	$\longrightarrow$ $K^{(I)}$ qua K-module	$\longrightarrow$ $L \otimes_K A$ qua pointed set	
adding some extra structure	$\longleftarrow$ choice of a basis	$\longleftarrow$ choice of a functorial A-module structure on V(A)	

So, a formal module differs from a formal variety by the functorial linear structure imposed on the pointed sets. More precisely, to define a formal module structure on a formal variety V, we have to choose two morphisms,

5.4 $$V \times V \to V \quad , \quad (x,y) \mapsto x + y \ ,$$

5.5 $$D \times V \to V \quad , \quad (\lambda,x) \mapsto \lambda x \quad ,$$

such that any V(A) becomes an A-module.

Then, for each formal variety W, the set $\mathfrak{M}(W,V)$ has a K-module structure, defined by using both above morphisms.

5.6 Far more interesting is the structure on the set $\mathfrak{M}(L^+,M^+)$ of the morphisms of a formal module into another. Indeed, we say that a morphism $f : L^+ \to M^+$ is <u>homogeneous of degree</u> $n \in \underline{\underline{P}}$, and we write $f \in \mathfrak{P}_n(L,M)$, iff

5.7 $$f(\lambda x) = \lambda^n f(x) \ .$$

In this relation, both sides stand for well defined morphisms from $D \times L^+$ to M^+. In other words, f_A has to be homogenous of degree n (in the usual sense), for each $A \in \underline{\underline{nil}}(K)$.

5.8 <u>Proposition. A set of morphisms of formal modules,</u> $\mathfrak{M}(L^+,M^+)$, <u>splits as the direct product of its homogeneous components,</u> $\mathfrak{P}_n(L,M)$, $n \in \underline{\underline{P}}$. <u>Every morphism</u> $f : L^+ \to M^+$ <u>can be written uniquely as</u>

$$f = \Sigma_{n \in \underline{\underline{P}}} f_n \quad , \quad f_n \in \mathfrak{P}_n(L,M) \quad ,$$

<u>the series converging in the</u> K-<u>module</u> $\mathfrak{M}(L^+,M^+)$ <u>for the order topology</u> (a fortiori for the simple topology).

5.9 In order to prove this, we choose an indexed basis in L, thereby identifying it with some $K^{(I)}$, then we write $f : D^{(I)} \to M^+$ in the form

$$f(x) = \Sigma_{\alpha \in \underline{\underline{N}}^{(I)}} \; c_\alpha x^\alpha \; ,$$

with $c_\alpha \in M$, following (5.2), and we put

$$f_n(x) = \Sigma_{\alpha \in \underline{\underline{N}}^{(I)}, |\alpha| = n} \; c_\alpha x^\alpha \; .$$

To check the unicity of the homogeneous components f_n of f, we can use free nilpotent algebras, as in (3.4).

5.10 Now, if $f : V \to W$ is a morphism of formal varieties, the homogenous components of f cannot be defined, but some relations are meaningful in the category of formal varieties, and can be expressed conveniently by using formal modules structures. Of such a kind is the relation $J_q f = J_q f'$ between $f, f' : V \to W$ (see 2.5). Namely, if we put formal module structures on V and W, thereby identifying them with L^+, M^+ respectively, then f and f' split as $f = \Sigma_n f_n$, $f' = \Sigma_n f'_n$, and $J_q f = J_q f'$ <u>is easily proved to be equivalent to</u>

$$f_n = f'_n \quad \text{for} \quad 1 \leq n \leq q \; .$$

For this reason, we introduce a new notation, by writing

5.11 $$J_q f = J_q f' \Longleftrightarrow f \equiv f' \quad \text{mod. deg. } q+1 \; ,$$

which means that f is congruent to f' modulo terms of degree

$\geqslant q+1$.

5.12 The order of a morphism $f : V \to W$, denoted by $\operatorname{ord}(f)$ is the symbol $+\infty$ iff f is the 0 morphism; if not, $\operatorname{ord}(f)$ is the greatest integer n for which

$$f \equiv 0 \quad \text{mod. deg. } n .$$

In other words, when $f = \Sigma_n f_n$ with homogeneous f_n, then $\operatorname{ord}(f)$ is the smallest n for which $f_n \neq 0$.

5.13 From what precedes, it is clear that, for any two formal varieties V,W , the set $\mathfrak{M}(V,W)$ of the morphisms $f : V \to W$ is the inverse limit of the sets $J_n(V,W)$ of the n-jets $J_n f$. This allows us to write

$$\mathfrak{M}(V,W) = \varprojlim J_n(V,W) = J_\infty(V,W) .$$

Furthermore all maps $J_\infty(V,W) \to J_n(V,W)$ are surjective (also, a fortiori, all maps $J_n(V,W) \to J_m(V,W)$ for $n > m$): this means only that there exist morphisms $f : L^+ \to M^+$ with prescribed homogeneous components up to a certain degree.

6. The functor $\mathfrak{T}$ (tangent)

6.1 For two K-modules, L,M and an integer $n \in \underline{\underline{P}}$, the K-module $\mathfrak{P}_n(L,M)$, defined in (5.6), is a set of morphisms of formal modules, $L^+ \to M^+$, not of maps $L \to M$ (although these notions may be identified if K is a $\underline{\underline{Q}}$-algebra). But, for $n=1$, $\mathfrak{P}_1(L,M)$ can always be identified to the module $\mathfrak{L}_K(L,M)$ of K-linear maps from L to M.

According to (5.10), we see that the first jet $J_1 f$ of a morphism of formal varieties, $f : V \to W$, is to be interpreted as a linear map of free modules. It remains to interpret the functors J_1 itself as a functor in free modules. We shall do it presently, introducing the sign $\mathfrak{T}$ instead of J_1.

Definition. Let V be a formal variety.

6.2 Any morphism $\gamma : D \to V$, from the "formal line" D, will be called a <u>curve</u> in V ; the set $\mathfrak{M}(D,V)$ will be denoted by $\mathfrak{C}(V)$.

6.3 The first jet $J_1\gamma$ of a curve $\gamma \in \mathfrak{C}(V)$ will be written $\mathfrak{T}\gamma$; we shall call it the <u>tangent vector</u> to V along γ (or the speed of γ). The set of tangent vector $\mathfrak{T}\gamma$, for $\gamma \in \mathfrak{C}(V)$, will be called the <u>tangent space</u> of V, and denoted by $\mathfrak{T}V$.

6.4 For a morphism of formal varieties, $f : V \to W$, the tangent map $\mathfrak{T}f : \mathfrak{T}V \to \mathfrak{T}W$ is defined by the formula

$$\mathfrak{T}f\cdot\mathfrak{T}\gamma = \mathfrak{T}(f\circ\gamma) \quad , \quad \text{for} \quad \gamma \in \mathfrak{C}(V) \ .$$

This is justified because $J_1(f\circ\gamma) = J_1 f\circ J_1\gamma$.

6.5 A curve in a formal module, $\gamma \in \mathfrak{C}(L^+)$ is written uniquely as a formal series,

$$\gamma(t) = \Sigma_{n\in \underline{\underline{P}}} \ a_n t^n \quad , \quad a_n \in L \ .$$

6.6 In this way the tangent vector $\mathfrak{T}\gamma$ becomes naturally identified with $a_1 \in L$. We transfer to $\mathfrak{T}L^+$ the K-linear structure of L. Then the tangent map $\mathfrak{T}f$ of any morphism $f : L^+ \to M^+$ becomes K-linear. From this, it follows that the <u>tangent space</u> $\mathfrak{T}V$ <u>of a formal variety</u> V <u>has a free</u> K-<u>module structure</u>, obtained here by imposing a formal module structure on V, but independent of the choice.

6.7 Let us repeat that, when V is a formal module L^+, or a model $D^{(I)}$, we can identify $\mathfrak{T}V$ to L, or to $K^{(I)}$, respectively, without ambiguity.

6.8 The tangent space $\mathfrak{T}V$ of a restricted product $\Pi^*_{i\in I} V_i$ is the direct sum of their tangent spaces: $\mathfrak{T}V = \oplus_{i\in I} \mathfrak{T}V_i$.

6.9 For a morphism $f : V \times V \to V$, we shall write

$$f(x,y) \equiv x + y \quad \text{mod. deg. } 2 ,$$

to mean that the tangent map $\mathfrak{T}f : \mathfrak{T}V \times \mathfrak{T}V \to \mathfrak{T}V$, is the addition of vectors.

6.10 The relation (6.9) holds iff both morphisms $x \mapsto f(x,0)$ and $x \mapsto f(0,x)$ have as tangent map the identity. A way to insure this is to assume that $x = f(x,0) = f(0,x)$. (In other words, f viewed as a binary operation has a two-sided neutral constant, which can only be 0).

In this way, the addition of tangent vectors is defined "intrinsically", that is without imposing any formal module structures. The multiplication of vectors by scalars in K may also be defined by the formula.

6.11 $$c\,\mathfrak{T}\gamma = \mathfrak{T}([c]\cdot\gamma) ,$$

where $[c]\cdot\gamma$ denotes the curve $t \to \gamma(ct)$, for $c \in K$.

6.12 _The dimension of a formal variety_ V _is defined as the rank of the free module_ $\mathfrak{T}V$ _over_ K.

Of course, the dimension of $D^{(I)}$ is the number of elements of I.

7. The composition lemma

7.1 Statement of the lemma. _Let_ U,V,W _be three formal varieties. Consider three morphisms_, $f,f' : U \to V$ _and_ $g : V \to W$, _such that_

$$f \equiv f' \equiv 0 \quad \text{mod. deg. } r ,$$

$$f \equiv f' \quad \text{mod. deg. } s ,$$

$$g \equiv 0 \quad \text{mod. deg. } t ,$$

where $r,s,t \in \underline{P}$ _and_ $r \leq s$. _Then_

$$g \circ f \equiv g \circ f' \quad \text{mod. deg. } (t-1)r+s .$$

This lemma contains useful information concerning morphisms of formal varieties, but we shall prove it for formal modules.

7.2 Our first step is to remark that the decomposition (5.8) of a morphism as the sum of its homogeneous components can be refined when a direct decomposition of the initial module is given. Note that $(L\times M)^+$ is naturally identified with $L^+ \times M^+$. Now, by choosing bases in L and in M, we can write a point in $L^+ \times M^+$ as a pair (x,y), where $x = (x_i)_{i\in I}$ and $y = (y_j)_{j\in J}$. By writing any monomial in (x,y) in the form $x^\alpha y^\beta$, where $\alpha \in \underline{\underline{N}}^{(I)}$, $\beta \in \underline{\underline{N}}^{(J)}$, and collecting in the general formula (5.2) the terms of the same degree in x and in y ($|\alpha| = m$ and $|\beta| = n$), we obtain a unique decomposition

7.3 $$f(x,y) = \Sigma_{m,n\in\underline{\underline{N}},\, m+n>0} \; f_{m,n}(x,y) \;, \quad \text{for any morphism}$$

$f : L^+ \times M^+ \to N^+$, where each $f_{m,n}$ is bihomogeneous, i.e. satisfies the identity

7.4 $$f_{m,n}(\lambda x, \mu y) = \lambda^m \mu^n f_{m,n}(x,y) \;.$$

Both sides of (7.4) denote morphisms $D^2 \times L^+ \times M^+ \to N^+$.

7.5 For instance, any morphism $f^*: L^+ \times D \to M^+$ can be written uniquely in the form

7.6 $$f^*(x,\lambda) = \Sigma_{m,n\in\underline{\underline{N}},m+n>0} \; \lambda^m f_{m,n}(x) \;, \quad \text{where} \quad f_{m,n} \in \mathfrak{P}_n(L,M) \,.$$

Furthermore, if f^* is defined by

$$f^*(x,\lambda) = f(\lambda x) \;,$$

where $f \in \mathfrak{M}(L^+,N^+)$, then the homogeneous decomposition $f = \Sigma_n f_n$ gives $f^*_{m,n} = 0$ if $m \neq n$ and $f^*_{n,m} = f_n$. From this we obtain the following characterization of the relation " $f \equiv 0$ mod. deg. r " (introduced as equivalent to $J_{r-1}f = 0$, or $\mathrm{ord}(f) \geqslant r$).

7.7 _A morphism_ $f : L^+ \to M^+$ _verifies_ $f \equiv 0$ mod. deg. r _iff there is a morphism_ $f^*: L^+ \times D \to M^+$ _such that_

$$f(\lambda x) = \lambda^r f^*(x,\lambda)$$

By considering the morphism $L^+ \times L^+ \to M^+$ given by $f(x+y) = \Sigma_{m,n} f_{m,n}(x,y)$ as in (7.3), we obtain still another characterization.

7.8 _A morphism_ $f : L^+ \to M^+$ _verifies_ $f \equiv 0$ mod. deg. r _iff the bihomogeneous components_ $f_{m,n}(x,y)$ _of_ $f(x+y)$ _are such that_ $m + n < r$ _implies_ $f_{m,n} = 0$.

Now let us prove the composition lemma (7.1), where we replace the formal varieties U,V,W by formal modules; this enables us to introduce the difference $h = f' - f$.

According to (7.8), we can write

7.9 $$g(x+y) = \Sigma_{m,n\in\underline{\underline{N}},m+n\geq t} \; g_{m,n}(x,y) \quad ,$$

with bihomogeneous $g_{m,n}$ as in (7.3). Note that $g_{m,o}(x,y)$ does not depend on y, and that

7.10 $$g(x) = \Sigma_{m\geq t} \; g_{m,o}(x,y)$$

Now, according to (7.7), the hypothesis concerning f and h is expressed by writing

$$f(\lambda x) = \lambda^r f^*(x,\lambda) \quad , \quad h(\lambda x) = \lambda^s h^*(x,\lambda) \; .$$

So we have, for any m,n ,

7.11 $$g_{m,n}(f(\lambda x), h(\lambda x)) = \lambda^{mr+ns} g_{m,n}(f^*(x,\lambda), h^*(x,\lambda)) .$$

What we have to prove is "$k \equiv 0$ mod. deg. $(t-1)r + s$" , where $k = g\circ f' - g\circ f$, and as, by (7.9 and 10),

7.12 $$k(x) = \Sigma_{m+n\geq t,n\geq 1} \; g_{m,n}(f(x), h(x)) \quad ,$$

it only remains to prove that $(t-1)r + s$ is the minimum of $mr + ns$, where $r \leq s$ and $m,n \in \underline{\underline{N}}$ are subject to $m+n \geq t$,

$n \geq 1$. This will complete the proof.

By putting, in (7.1), $f' = 0$ and $r = s$, we obtain the following special case (easy to prove directly).

7.13 Let $f : U \to V$ and $g : V \to W$ be two morphisms of formal varieties. Then $\operatorname{ord}(g \circ f) \geq \operatorname{ord}(g)\ \operatorname{ord}(f)$.

8. The isomorphism theorem

8.1 Theorem. A morphism of formal varieties, $f : V \to W$, is an isomorphism iff its tangent map $\mathfrak{T}f : \mathfrak{T}V \to \mathfrak{T}W$, is an isomorphism of modules.

8.2 We prove the theorem for formal modules.

The first step is a reduction to the case where $W = V$ and $\mathfrak{T}f = \mathrm{Id}$. For this, we have only to find an isomorphism $g : W \to V$, such that $\mathfrak{T}(g \circ f) = \mathfrak{T}g \circ \mathfrak{T}f$ be the identity of $\mathfrak{T}V$; this is easy by taking g linear (homogeneous of degree 1).

8.3 Now, a morphism $f : L^+ \to L^+$, with $\mathfrak{T}f = \mathrm{Id}$, can be written in the form

$$f(x) = x - g(x) \ ,$$

8.4 where

$$g \equiv 0 \quad \text{mod. deg. } 2 \ .$$

We put

8.5
$$F(x,y) = x + g(y) \ ,$$

and the equation $f \circ h = \mathrm{Id}$ (where h is the right inverse to f we are looking for) can be rewritten as

8.6
$$h(x) = F(x,h(x)) \ .$$

The composition lemma (7.1) shows that, for any $h,h' \in \mathfrak{M}(L^+,L^+)$, the relation $h \equiv h'$ mod. deg. r implies $F(x,h(x)) \equiv F(x,h'(x))$ mod. deg. $(r+1)$. This shows the unicity of the solution h of equation (8.6), and h is obtained as the limit (for the order topology) of an iteration sequence $h_o, h_1, \ldots, h_{n+1}(x) = F(x, h_n(x))$, starting with any h_o.

This is only the formal part of the classical proof of the local inversion theorem for analytic manifolds.

9. The fibration $J_q \to J_{q-1}$

9.1 The functor J_1 is identified by canonical isomorphisms to the functor $\mathfrak{T}$ (see 6). Here we shall discuss a property of the functor J_q for $q > 1$ (see 2.4).

Let $f, f': V \to W$ be two morphisms of formal varieties, having the same (q-1)-th jet: $J_{q-1}f = J_{q-1}f'$. We will show presently that there is an element of $\mathfrak{P}_q(\mathfrak{T}V, \mathfrak{T}W)$, called _the difference of_ f' _and_ f _in degree_ q and denoted by

9.2 $$\mathrm{dif}_q(f', f) .$$

In the case where $V = L^+$ _and_ $W = M^+$ _are given as formal modules, we have the homogeneous decompositions_ $f = \Sigma_n f_n$, $f' = \Sigma_n f'_n$ of f and f' ; _the hypothesis_ $J_{q-1}f = J_{q-1}f'$ _means that_ $f_n = f'_n$ for $n < q$. _Then, by identifying_ $\mathfrak{T}V, \mathfrak{T}W$ _with_ L, M _respectively_ (6.6), _we have_

9.3 $$f'_q - f_q = \mathrm{dif}_q(f', f) \in \mathfrak{P}_q(\mathfrak{T}V, \mathfrak{T}W) .$$

The point to check is that the difference $\mathrm{dif}_q(f', f)$ is well defined by the above formula, i.e. does not depend on the formal module structures imposed on formal varieties.

9.4 _Lemma. Let_ $f, f': L^+ \to M^+$ _and_ $g, g' \in M^+ \to N^+$ _be morphisms of formal modules, such that_ $J_{q-1}f = J_{q-1}f'$ _and_ $J_{q-1}g = J_{q-1}g'$ _for some_ $q > 1$. _Then we have_

9.5 $$\mathrm{dif}_q(g' \circ f', g \circ f) = \mathrm{dif}_q(g', g) \circ f_1 + g_1 \circ \mathrm{dif}_q(f', f),$$

where dif_q *is defined as in* (9.3), *and* f_n, etc. *denote the homogeneous component of degree* n *of* f, etc.

Proof. The sum in the right side of (9.5) corresponds to:

$$g'\circ f' - g\circ f = (g'\circ f' - g\circ f') + (g\circ f' - g\circ f) .$$

Now $g'\circ f' - g\circ f' = (g'-g)\circ f'$, and, as $g'-g \equiv \mathrm{dif}_q(g',g)$ mod. deg. $(q+1)$ and $f' \equiv f$, mod. deg. 2 , the composition lemma (7.1) gives us

9.6 $$g'\circ f' - g\circ f' \equiv \mathrm{dif}_q(g',g)\circ f_1 \text{ . mod. deg. } (q+1).$$

Another direct application of the composition lemma gives us

9.7 $$g\circ f' - g\circ f \equiv g_1\circ \mathrm{dif}_q(f',f) \text{ mod. deg. } (q+1),$$ which completes the proof.

9.8 Let $\varphi : L \to L_1$ and $\psi : M \to M_1$ be two linear isomorphisms of free K-modules, which can also be viewed as isomorphisms $\varphi : L^+ \to L_1^+$ and $\psi : M^+ \to M_1^+$. Let $f,f' \in \mathfrak{M}(L^+,M^+)$ and $f_1, f_1' \in \mathfrak{M}(L_1^+,M_1^+)$ be such that $J_{q-1}f = J_{q-1}f'$, $\psi\circ f = f_1\circ\varphi$, $\psi\circ f' = f_1'\circ\varphi$. Then

$$\psi\circ \mathrm{dif}_q(f',f) = \mathrm{dif}_q(f_1',f_1)\circ\varphi .$$

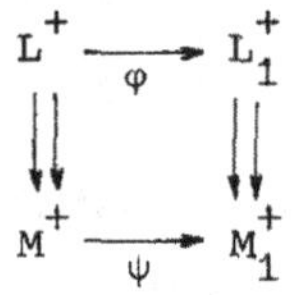

This assertion follows from lemma (9.5), and legitimates the definition (9.3) of dif_q for morphisms of formal varieties (not only formal modules).

9.9 *Let* $E = \mathfrak{M}(V,W)$ *be the set of morphisms of two formal varieties*. *Then*, *for any* $q > 1$, J_qE *is an affine bundle over* $J_{q-1}E$. *More precisely*, *the inverse image of any point of* $J_{q-1}E$, *for the canonical map* $J_qE \to J_{q-1}E$, *is a principal homogeneous space over the* K-*module* $\mathfrak{P}_q(\mathfrak{T}V,\mathfrak{T}W)$. *If* $f,f' \in E$, $J_{q-1}f = J_{q-1}f'$, *then the element of* $\mathfrak{P}_q(\mathfrak{T}V,\mathfrak{T}W)$ *mapping* J_qf *on* J_qf' *is denoted by* $\mathrm{dif}_q(f',f)$.

Let U,V,W *be formal varieties and* $f,f': U \to V$, $g,g': V \to W$ *be morphisms such that* $J_{q-1}f = J_{q-1}f'$, $J_{q-1}g = J_{q-1}g'$ $(q > 1)$. *Then*

9.10 $$\mathrm{dif}_q(g'\circ f', g\circ f) = \mathrm{dif}_q(g',g)\circ \mathfrak{T}f + \mathfrak{T}g\circ \mathrm{dif}_q(f',f) \ .$$

The statements (9.9 and 10) are just "intrinsic" reformulations of the definition of dif_q and of lemma (9.4). Note that we identify the tangent map $\mathfrak{T}f$ of a morphism $f : V \to W$ with an element of $\mathfrak{P}_1(\mathfrak{T}V,\mathfrak{T}W)$, and that the definition of homogeneous morphisms (5.7) implies immediately

9.11 $$g\circ f \in \mathfrak{P}_{mn}(L,N) \quad \textit{for} \quad f \in \mathfrak{P}_m(L,M) \quad \textit{and} \quad g \in \mathfrak{P}_n(M,N).$$

9.12 The set of morphisms $f : V \to W$ of two given formal varieties such that $J_{q-1}f = 0$ can be identified with $\mathfrak{P}_q(\mathfrak{T}V,\mathfrak{T}W)$, because there is an origin in the corresponding affine space, namely the 0 q-jet. If $q = 2$, the element of $\mathfrak{P}_2(\mathfrak{T}V,\mathfrak{T}W)$, corresponding to a morphism $f : V \to W$ with $\mathfrak{T}f = 0$, is called the *hessian* of f, so that $\mathrm{Hess}\, f = \mathrm{dif}_2(f,0)$.

If $f : L^+ \to M^+$ is such that $J_{q-1}f = 0$, then $f(\lambda x) = \lambda^q f^*(x,\lambda)$ as in (7.7), and we have

$$\mathrm{dif}_q(f,0) = f^*(x,0) \ .$$

10. *The functor* $\mathfrak{C}$ *(curves)*

10.1 We have already introduced the set $\mathfrak{C}(V) = \mathfrak{M}(D,V)$ of curves in a formal variety V (6.2). The functor $\mathfrak{C}$ is defined in the usual way, i.e. by associating to any morphism $f : V \to W$ of formal varieties the map $\mathbb{C}(f): \mathbb{C}(V) \to \mathbb{C}(W)$, defined by

10.2 $$\mathbb{C}(f)(\gamma) = f\circ\gamma \quad , \quad \text{for} \quad \gamma \in \mathbb{C}(V) \ .$$

10.3 The curves lemma. The functor $\mathbb{C}$ is faithful. In other words, if $f, f' : V \to W$ are two morphisms of formal varieties such that $f\circ\gamma = f'\circ\gamma$ for every $\gamma \in \mathbb{C}(V)$, then $f = f'$.

Proof. An easy reduction shows that it suffices to prove the lemma when $V = D^n$ (for some integer n), $W = D$ and $f' = 0$. So let $f \in \mathfrak{M}(D^n, D)$ be given by

10.4 $$f(x) = \Sigma_{\alpha\in\underline{\underline{N}}^n} \, c_\alpha x^\alpha \ , \quad c_\alpha \in K, \ c_o = 0 \ .$$

If $f \neq 0$, then we shall presently prove that $f\circ\gamma \neq 0$, where the curve $\gamma : D \to D^n$ is defined by

10.5 $$\gamma(t) = (t^{d^{n-1}}, t^{d^{n-2}}, \dots, t^d, t) \ ,$$

for sufficiently large $d \in \underline{\underline{P}}$.

Indeed, let us well-order lexicographically the multidegrees $\alpha \in \underline{\underline{N}}^n$, and let β be the least one for which the coefficient c_β in (10.4) does not vanish. When we substitute $\gamma(t)$ for x in a monomial $c_\alpha x^\alpha$, we obtain the monomial in t with coefficient c_α and degree

10.6 $$\alpha_1 d^{n-1} + \alpha_2 d^{n-2} + \dots + \alpha_n \ .$$

Now we have

10.7 $$\Sigma_{1\leq i\leq n} \, \alpha_i d^{n-i} > \Sigma_{1\leq i\leq n} \, \beta_i d^{n-i} \ ,$$

provided that $\alpha > \beta$ (for the lexicographic order), and

10.8 $$d > \max_{1\leq i\leq n} \beta_i \ ,$$

because the first non-vanishing difference $\alpha_i - \beta_i$, say $\alpha_j - \beta_j$, is ≥ 1, so that (10.7) follows from

$$d^{n-j} > \Sigma_{j+1\leq i\leq n} \, \beta_i d^{n-i} \ ,$$

which is ensured by the condition (10.8). The order of $\gamma \circ f$ is finite, therefore $\gamma \circ f \neq 0$.

It is easy to characterize the natural applications (or functorial morphisms) $x : \mathfrak{C} \to \mathfrak{C}$. For every formal variety V, one should have a map $x_V: \mathfrak{C}(V) \to \mathfrak{C}(V)$, with the condition that

10.9 $$x_W \cdot (f \circ \gamma) = f \circ (x_V \cdot \gamma) ,$$

for any morphism $f : V \to W$. By replacing, in (10.9), V by D, W by V, f by γ and γ by Id_D, we obtain $x_V \cdot \gamma = \gamma \circ \varphi$, where $\varphi = x_D \cdot Id_D$. For every morphism $\varphi \in \mathfrak{C}(D)$, we define the <u>composition "operator"</u> $comp(\varphi)$ by putting

10.10 $$comp(\varphi) \cdot \gamma = \gamma \circ \varphi , \text{ for any curve } \gamma .$$

As φ has been shifted from the right to the left in the above formula, we have

10.11 $$comp(\varphi' \circ \varphi) = comp(\varphi)\ comp(\varphi') , \text{ for any}$$

$\varphi, \varphi' \in \mathfrak{C}(D)$, where, on the right side, juxtaposition of operators denotes their composition (when they act on the left).

Some composition operators will receive special notations. Namely:

10.12 $$V_n = comp(\varphi), \text{ for } \varphi(t) = t^n, n \in \underline{\underline{P}} ;$$

10.13 $$[c] = comp(\varphi), \text{ for } \varphi(t) = ct, c \in K \text{ (as in 6.11)}$$

The following relations are immediate consequences of (10.11).

10.14 $$[1_K] = V_1 = Id \text{ (the identity operator) } ;$$

10.15 $$[cc'] = [c][c'], \text{ for any } c, c' \in K ;$$

10.16 $$V_m V_n = V_{mn} , \text{ for any } m, n \in \underline{\underline{P}} ;$$

10.17 $$[c]V_n = V_n[c^n] , \text{ for any } c \in K, n \in \underline{\underline{P}} .$$

10.18 Definition. An indexed set of curves $(\gamma_i)_{i\in I}$ in a formal variety V will be called a basic set iff their tangent vectors $(\mathfrak{T}\gamma_i)_{i\in I}$ are a basis of the free module $\mathfrak{T}V$.

11. Changes of rings

11.1 Until now, the basic ring K remained unchanged, and mostly implicit. A change of basic rings will presently be defined for any ring homomorphism, $\varphi : K \to K'$.

To every model over K, which we denote now by $D_K^{(I)}$, there corresponds the model over K' with the same indexing set, namely $D_{K'}^{(I)}$. To every morphism $f : D_K^{(I)} \to D_K^{(J)}$, there corresponds the morphism

$$\varphi_* f : D_{K'}^{(I)} \to D_{K'}^{(J)} ,$$

obtained by applying the homomorphism φ to every coefficient of the formal series defining f. More precisely, if $f = (f_j)_{j\in J}$, $f_j(x) = \Sigma_\alpha c_{j,\alpha} x^\alpha$, then

11.2 $$(\varphi_* f)_j = \Sigma_\alpha \varphi(c_{j,\alpha})x^\alpha , \; j \in J, \; \alpha \in \underline{\underline{N}}^{(I)} .$$

We have just defined the functor φ_* , from the category of models over K to the category of models over K'. Note that

11.3 $$(\psi\circ\varphi)_* = \psi_* \circ \varphi_* ,$$

for composable ring homomorphisms φ, ψ .

Now, we want to extend the functor φ_* to the category of formal varieties over K.

Let V be a formal variety over K. We have to define $\varphi_* V$ as a formal variety over K', i.e. to define the pointed set $(\varphi_* V)(A')$ for any $A' \in \underline{\underline{\underline{nil}}}(K')$. Let us denote by $\varphi^* A' \in \underline{\underline{\underline{nil}}}(K)$ the K-algebra with the same underlying ring as A', multiplication by scalars from K being defined by

11.4 $$c\,a' = \varphi(c)a' \quad , \quad \text{for} \quad c \in K,\ a' \in A' \ .$$

Then we put

11.5 $$(\varphi_* V)(A') = V(\varphi^* A') \ ,$$

and, to any morphism $f : V \to W$ over K, there corresponds the morphism $\varphi_* f : \varphi_* V \to \varphi_* W$ over K', defined by

11.6 $$(\varphi_* f)_{A'} = f_{(\varphi^* A')} \ , \quad \text{for} \quad A' \in \underline{\underline{\underline{nil}}}(K') \ .$$

One checks easily that, when applied to models, this general definition of the functor φ_* is consistent with the former one (11.1).

The functor $\varphi*$ commutes to products: for any set $(V_i)_{i \in I}$ of formal varieties over K,

11.7 $$\varphi_*(\Pi_{i\in I} V_i) = \Pi^*_{i\in I} \varphi * V_i \ .$$

The functor $\varphi*$ can be applied to jets (see 2.4), because any relations $J_n f = J_n f'$ implies $J_n \varphi_* f = J_n \varphi * f'$ (or, categorically, because φ^* sends $\underline{\underline{\underline{nil}}}(K',n)$ into $\underline{\underline{\underline{nil}}}(K,n)$ for any n), and

11.8 $$J_n \varphi_* = \varphi_* J_n \ ,$$

both sides beeing applied to a formal variety over K or to a morphism over K.

As $\mathfrak{T} = J_1$ up to isomorphism, we have a map

11.9 $$\varphi_* : \mathfrak{T}V \to \mathfrak{T}\varphi_* V \ ,$$

for any formal variety V over K, defined by

11.10 $$\varphi_*(\mathfrak{T}\gamma) = \mathfrak{T}(\varphi_* \gamma) \quad \text{for} \quad \gamma \in \mathfrak{C}(V) \ .$$

Note also that

11.11 $$\varphi_*(V_n \cdot \gamma) = V_n \cdot (\varphi_* \gamma) \quad , \quad n \in \underline{\underline{P}} \ ;$$

11.12 $$\varphi_*([c]\cdot\gamma) = [\varphi(c)]\cdot\varphi_*\gamma, \ c \in K \ .$$

11.13 When applying φ_* to a formal module L^+ over K, we can identify $\varphi_* L^+$ with the formal module L'^+ over K', where $L' = K' \otimes_\varphi L$. Consistence with the general definition of φ_* is expressed by the following identifications:

$$(\varphi_* L^+)(A') = (\varphi^* A') \otimes_K L = A' \otimes_\varphi L = A' \otimes_{K'} (K' \otimes_\varphi L) = A' \otimes_{K'} L' .$$

11.14 If $(\gamma_i)_{i \in I}$ is a basic set of curves in a formal variety V over K, then $(\varphi_* \gamma_i)_{i \in I}$ is a basic set of curves in $\varphi_* V$ (see 10.18).

11.15 The tangent space $\mathfrak{T}(\varphi_* V)$ of $\varphi_* V$ can be identified with the tensor product $K' \otimes_\varphi \mathfrak{T}V$, the map (11.9) becoming $v \mapsto 1 \otimes v$, for $v \in \mathfrak{T}V$.

The following important proposition is evident when one looks at models (11.1).

11.16 _Let_ $\varphi : K \to K'$ _be an injective (resp. surjective) ring homomorphism and_ V, W _two formal varieties over_ K. _Then the map_ $f \mapsto \varphi_* f$ _of_ $\mathfrak{M}(V,W)$ _into_ $\mathfrak{M}(\varphi_* V, \varphi_* W)$ _is injective (resp. surjective)_.

11.17 Formal varieties and morphisms over $\underline{\underline{Z}}$ deserve a special mention, because, for any basic ring K, there is a _unique_ ring homomoprhism $\varphi : \underline{\underline{Z}} \to K$, so that the changes of rings, starting from $\underline{\underline{Z}}$, may be said to be automatic.

For instance, the morphism $\mathrm{sym}_n : D^n \to D^n$ is defined by the elementary symmetric polynomials:

11.18 $$\mathrm{sym}_n(x_1, x_2, \ldots, x_n) = (\Sigma_i x_i, \Sigma_{i<j} x_i x_j, \ldots, x_1 \ldots x_n)$$

11.19 _The symmetric morphism theorem. Let_ $f : D^n \to V$ _be a symmetric morphism, i.e. a morphism verifying_

$$f(x_1, \ldots, x_n) = f(x_{\sigma(1)}, \ldots, x_{\sigma(n)})$$

<u>for any permutation</u> σ <u>of the indices</u> $1,\ldots,n$. <u>Then there is a unique morphism</u> $g : D^n \to V$, <u>such that</u>

11.20 $$f = g \circ \mathrm{sym}_n \, .$$

By taking a formal module structure on V, this theorem reduces to the classical one concerning symmetric polynomials. As pointed out (11.17), the basic ring K may remain undefined.

CHAPTER II

FORMAL GROUPS AND BUDS

1. <u>Groups in categories</u>

1.1 There are many equivalent ways to define <u>groups</u>, <u>or commutative groups</u>, in a <u>category with finite products</u>. Anyhow, a group is an object G in the category, with some extra structure which admits alternative descriptions.

1.2 We can define a structure of group (resp. of commutative group) on G by giving, for each <u>word-function</u> in n variables from group theory (resp. commutative group theory) the corresponding <u>word-morphism</u>, $G^n \to G$, in the category. <u>Identities relating word-functions are assumed to hold for word-morphisms</u>.

1.3 All word-morphisms can be derived from the morphism $G^2 \to G$, corresponding to the group operation. Let us state the required properties of f.

<u>Associativity</u> of f means commutativity of the following diagram.

1.4

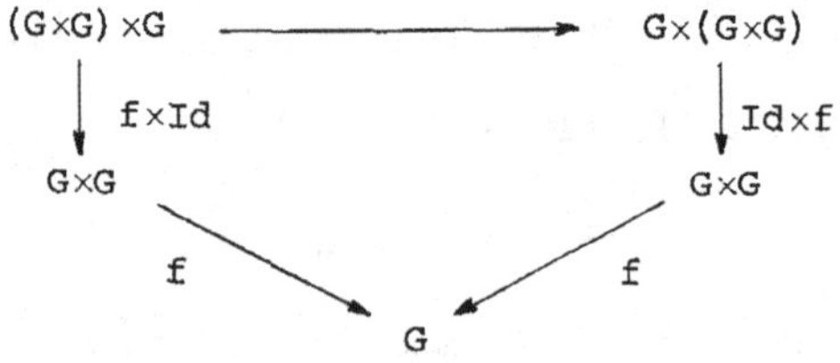

The others properties required for f to define a group structure in the category may be summed up as follows.

1.5 <u>The morphism</u> $G^2 \to G^2$, $(x,y) \mapsto (x,f(x,y))$, <u>is an isomorphism</u>.

This corresponds to the axiomatic definition of a group as a monoid (associative system), where all left multiplications are bijective.

<u>Commutativity</u> of f means commutativity of the following diagram

1.6

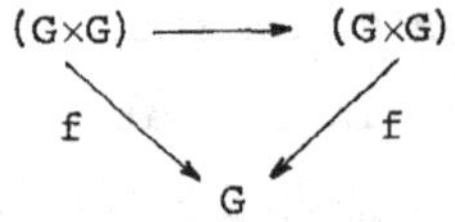

where the horizontal arrow denotes the involution $(x,y) \mapsto (y,x)$.

1.7 An alternative way of defining a group structure on G is to impose an (ordinary) group structure on the set of morphisms Mor(V,G) for any object V in the category, in such a way that we obtain a <u>contravariant functor in groups</u>. More precisely, for any morphism $u : V \to W$ in the category, the map $g \mapsto g \circ u$ from Mor(W,G) into Mor(V,G) is required to be a group homomorphism.

1.8 We shall use this definition in two steps: first there will be defined a contravariant functor in groups, $V \mapsto \Gamma(V)$, as above. Then it will be proved that this functor is <u>representable</u> in the category, i.e. that $\Gamma(V)$ *can be functorially identified with* Mor(V,G) *for some object* G *in the category* (G is defined up to canonic isomorphism in the category).

A group *homomorphism* in a category is a morphism, $u : G \to G'$, which "commutes to the word-morphisms" or (which amounts to the same) commutes to group operations; i.e. the following diagram is commutative

1.9

$$\begin{array}{ccc} G\times G & \xrightarrow{u\times u} & G'\times G' \\ \downarrow f & & \downarrow f' \\ G & \xrightarrow{u} & G' \end{array}$$

where f, f' denote the group morphisms of G, G' respectively.

1.10 Alternatively, we can define a group homomorphism $u : G \to G'$ as a morphism such that, for any object V, the map $\mathrm{Mor}(V,G) \to \mathrm{Mor}(V,G')$, $g \mapsto u \circ g$, is a group homomorphism.

1.11 The set of homomorphisms from G to G' will be denoted by Hom(G,G'). If G and G' are commutative, then Hom(G,G') has a *group structure*. If G is commutative, the set Hom(G,G) = End(G) has a *ring structure*.

2. *Group laws and formal groups*

2.1 *A convention of language*. As we shall remain almost exclusively interested in *commutative* groups, *we shall just say* "*group*" instead of "commutative group". When we come to discuss some points concerning "non necessarily commutative groups", we shall expressly say so.

2.2 *Definitions*. *A formal group is a group in the category of formal varieties. A group law is a group in the category of* models. (Of course, there is some given basic ring).

2.3 A formal group is a formal variety G, given with a morphism $\mu : G^2 \to G$. Apart from associativity and commutativity (see I.4.5 and diagrams 1.4, 1.9), μ is assumed to verify the condition (1.5), which can be replaced by the following one:

2.4 $$\mu(x,y) \equiv x + y \quad \text{mod. deg. } 2 .$$

Indeed, this condition follows from the existence of the neutral constant 0 in a formal group (see I.6.10). Conversely, condition (2.4) implies (1.5), according to the isomorphism theorem.

Every coordinate system $f : D^{(I)} \to G$ on a formal group G (see I.4.1) defines a group law μ_f on the model $D^{(I)}$, by the formula

2.5 $$f(\mu_f(x,y)) = \mu(f(x),f(y)) ,$$

or alternatively by

2.6 $$\mu_f(x,y) = f^{-1}(\mu(f(x),f(y))) .$$

2.7 By choosing once for all the indexing set I, and letting f range over the whole set of isomorphisms $D^{(I)} \to G$, the group laws μ_f make up an isomorphy class in the set of all group laws on $D^{(I)}$.

We can say that formal groups versus group laws appear as an incentive to study group laws up to isomorphism (which, unfortunately, cannot be achieved), or at least not to let wanton use of indices obscure clear concepts.

2.8 For a formal group G and any formal variety V, the set of morphisms $\mathfrak{M}(V,G)$ is a group (see 1.7), which we shall write additively. The group structure on $\mathfrak{M}(V,G)$ is compatible with both its simple and its order topology (see I.2.6, I.2.7); let us recall that there is no difference between those topologies if V is finite-dimensional (see I.3.7).

2.9 For two formal groups G,G', we have to distinguish the group $\mathfrak{M}(G,G')$ of all morphisms $G \to G'$ (where the group structure of G is disregarded) from its subgroup $\mathrm{Hom}(G,G')$, made up by all the group-homomorphisms $G \to G'$ (see 1.9). Note that the subgroup $\mathrm{Hom}(G,G')$ is closed in $\mathfrak{M}(G,G')$, both for its simple and its order topologies.

2.10 The category of formal groups is defined by taking the sets of group-homomorphisms as sets of morphisms. This category contains finite and infinite direct sums (corresponding to restricted products).

2.11 Change of rings. Let $\varphi : K \to K'$ be a ring homomorphism and G, $\mu : G^2 \to G$, be a formal group over K. Then $\varphi_* G$, endowed with $\varphi_* \mu$, is a formal group over K'. Moreover, for any formal variety V over K, the map $f \mapsto \varphi_* f$, of $\mathfrak{M}(V,G)$ into $\mathfrak{M}(\varphi_* V, \varphi_* G)$, is a topological-group homomorphism. If $f : G \to G'$ is a formal-group homomorphism over K, then $\varphi_* f : \varphi_* G \to \varphi_* G'$ is a formal-group homomorphism over K'. In other words, φ_* defines a functor from formal groups over K to formal groups over K'.

All this appears rather obvious when looking at the definition of φ_* for group laws (see I.11.1). Categorical proofs are obtained by applying φ_* to some commutative diagrams, such as (1.4), (1.6), (1.9), where some morphisms are defined over $\underline{\underline{Z}}$ (see I.11.17).

2.12 A one-dimensional group law is just a formal series

$f : D \times D \to D$ which verifies the axioms of group laws. _We shall denote by_ $\underline{\underline{G}}_a$ _the additive group law_

2.13 $$x + y .$$

We define the multiplicative group law by

2.14 $$x + y - xy ,$$

and denote it by $\underline{\underline{G}}_m$. The minus sign in (2.14) may look strange (one would rather expect $x + y + xy$), but it will be justified later as a way of saving many "$(-1)^n$".

Both $\underline{\underline{G}}_a$ and $\underline{\underline{G}}_m$ are defined over $\underline{\underline{Z}}$, that is over any basic ring K. A word-function (see 1.2),

2.15 $$\Sigma_{1\leq i\leq n}\, c_i x_i \qquad (c_i \in \underline{\underline{Z}}) ,$$

from group theory corresponds to a word-morphism written just as (2.15), viewed as a formal series, in $\underline{\underline{G}}_a$; whereas in $\underline{\underline{G}}_m$ it corresponds to the series $h(x_1,\ldots,x_n)$ defined by

2.16 $$1 - h(x_1,\ldots,x_n) = \Pi_{1\leq i\leq n}\, (1-x_i)^{c_i}$$

where

2.17 $$(1-x)^c = 1 - \Sigma_{n\in\underline{\underline{P}}}\, (-1)^{n-1} \binom{c}{n} x^n .$$

The groups of curves $\mathfrak{C}_K(\underline{\underline{G}}_a)$, $\mathfrak{C}_K(\underline{\underline{G}}_m)$ depend on the basic ring K. An element $\gamma : D \to D$ in either is written as a formal series

2.18 $$\gamma(t) = \Sigma_{n\in\underline{\underline{P}}}\, a_n t^n \qquad (a_n \in K) .$$

Let γ, γ', γ'': $D \to D$ be defined by the sequences (a_n), (a'_n), (a''_n) in K, $n \in \underline{\underline{P}}$. Then

2.19 $$\gamma + \gamma' = \gamma'' \quad \underline{\text{in}} \quad \mathfrak{C}(\underline{\underline{G}}_a)$$

means

2.19.bis $$a''_n = a_n + a'_n \qquad \text{for any } n \in \underline{\underline{P}} ;$$

whereas

2.20 $$\gamma + \gamma' = \gamma'' \qquad \underline{\text{in}} \qquad \mathfrak{C}(\underline{\underline{G}}_m)$$

means

2.20.bis $a''_n = a_n + a'_n - \Sigma_{1\leq i<n} a_i a'_{n-i}$, for any $n \in \underline{\underline{P}}$.

2.21 Generally, we define the underlying additive group of any model $D^{(I)}$ or of any formal module L^+ by keeping its structure of formal variety and its additive law, written as (2.13) but with arbitrary dimension, and forgetting the rest.

3. The $\underline{\underline{Q}}$ theorem, the lift theorem, and the general existence of ghosts

In this section, we state and discuss some theorems, which are fundamental for the theory of formal groups.

3.1 The $\underline{\underline{Q}}$ theorem. Let K be a $\underline{\underline{Q}}$-algebra and G,G' two formal groups over K. Then, to any K-linear map $u : \mathfrak{T}G \to \mathfrak{T}G'$, there is one and only one formal-group homomorphism $f : G \to G'$ such that $\mathfrak{T}f = u$.

3.2 In other words, the category of formal groups over K (see 2.10) is equivalent to the category of the free K-modules, the functor $\mathfrak{T}$ beeing fully faithful.

3.3 Up to isomorphism, there is only one group law for any dimension: the additive one (see 2.21).

3.4 For any formal group G over K (a $\underline{\underline{Q}}$-algebra), there is a well defined pair of formal-group isomorphisms, inverse to each other, namely

3.5 $$\log_G : G \to (\mathfrak{T}G)^+ ,$$

3.6 $$\exp_G : (\mathfrak{T}G)^+ \to G ,$$

defined by the condition that their tangent maps are the identity map of $\mathfrak{T}G$ (identified to the tangent space of $(\mathfrak{T}G)^+$, as in I.6.7).

3.7 As an example, if G is the multiplicative group $\underline{\underline{G}}_m$ over $\underline{\underline{Q}}$ (see 2.14), then the series (3.5) and (3.6) become respectively

$$\Sigma_{n\in\underline{\underline{P}}} \frac{x^n}{n} \quad \text{and} \quad \Sigma_{n\in\underline{\underline{P}}} (-1)^{n-1} \frac{x^n}{n!} \ .$$

3.8 The lift theorem. Let $\varphi : K_o \to K$ be a surjective ring homomorphism, and $f : D^{(I)} \times D^{(I)} \to D^{(I)}$ be a group law over K. Then there is a group law $f_o : D^{(I)} \quad D^{(I)} \to D^{(I)}$ over K_o, such that $f = \varphi_* f_o$.

We stated the lift theorem for group laws, but it holds for formal groups: any formal group G over K is isomorphic to $\varphi_* G_o$, for a suitable formal group G_o over K_o.

3.9 Combining the $\underline{\underline{Q}}$ theorem with the lift theorem, we obtain the following story.

Let K be any ring. We can choose a torsion-free ring K_o and a surjective ring-homomorphism $\varphi : K_o \to K$ (for instance by taking for K a suitable ring of polynomials over $\underline{\underline{Z}}$). We denote by $\psi : K_o \to K$, the canonic embedding of K_o into $K_1 = \underline{\underline{Q}} \otimes_{\underline{\underline{Z}}} K_o$. Let, for some indexing set I, $\omega : D_{K_1}^{(I)} \to D_{K_1}^{(I)}$ be an isomorphism such that $\omega(x) \equiv x$ mod.deg. 2 (i.e. $\mathfrak{T}\omega = \mathrm{Id}$). Let f_1 be the group law over K_1 defined by

$\omega(f_1(x,y)) = \omega(x) + \omega(y)$, where the sum on the right side refers to the additive group of the model $D^{(I)}$.

It may happen that the formal series which make up f_1 have their coefficients in ψK_o. This is expressed by writing $f_1 = \psi_* f_o$, where f_o is a group law over K_o . In this case, we can reduce, modulo some ideal in K_o , the coefficients of the formal series in f_o: in other words, we can consider the group law $f = \varphi_* f_o$ over K. Every group law f over K can be obtained in this way, for the lift theorem ensures the existence of f_o and the $\underline{\underline{Q}}$ theorem ensures the existence of ω. If we think of x as of a point in some space over K, then $\omega(x)$ is a ghost, by definition. See, later (III.4.9)

4. Buds and extensions

4.1 For any $n \in \underline{\underline{P}}$, we define a n-bud as a group in a category of n-jets (see I.2.4), in the standard way (1.2).

Practically, a n-bud is defined by a morphism $f : V \times V \to V$ (where V may be a model, a formal module, or a formal variety), verifying

4.2 $$f(x,y) \equiv x + y \qquad \text{mod.deg. } 2 ,$$

4.3 $$f(x,y) \equiv f(y,x) \qquad \text{mod.deg. } (n+1),$$

4.4 $$f(f(x,y),z) \equiv f(x,f(y,z)) \qquad \text{mod.deg. } (n+1) .$$

Any morphism $f' : V \times V \to V$, with $J_n f = J_n f'$, defines the same n-bud as f; we say that f is a representative of the n-bud. In the case where V has a formal module structure, we can take representative of n-buds with vanishing homogeneous components of degree $> n$.

There is essentially one 1-bud for each dimension, denoted by $x+y$. A 2-bud is defined by $x + y + a(x,y)$, where a is bilinear (i.e. of total degree 2 and such that $a(0,y)=a(x,0)=0$) and symmetric (i.e. $a(x,y) = a(y,x)$).

4.5 Homomorphisms of n-buds are defined as n-jets of morphisms, and by morphisms. What was said in (2.11) concerning changes of basic rings applies to n-buds and n-buds homomorphisms.

4.6 Statement of the extension theorem. Every n-bud, for any $n \in \underline{\underline{P}}$, over any basic ring, has a group law as a representative.

4.7 The $\mathbb{Q}$ theorem (3.1), the lift theorem (3.8) and the extension theorem (4.6) will be proved by an induction on n, using n-buds. Here we shall only formulate the induction hypotheses, depending on the integer q. The ring homomorphisms $\varphi : K_o \to K$ and $\psi : K_o \to K_1$ are as in the ghost story (3.9), i.e. K_o is torsion-free, φ is surjective, $K_1 = \mathbb{Q} \otimes_{\mathbb{Z}} K_o$, and we shall identify K_o with its image ψK_o, so that K_o becomes a subring of K_1, with inclusion map $\psi : K_o \to K_1$. The indexing set I is fixed, and we consider buds on the model $D^{(I)}$.

4.8(A_q) For any q-bud over K_1, represented by the morphism f, there is an isomorphism $\omega : D^{(I)}_{K_1} \to D^{(I)}_{K_1}$, with $\omega(x) \equiv x$ mod. deg. 2 and

$$\omega(f(x,y)) \equiv \omega(x) + \omega(y) \quad \text{mod. deg. } (q+1) ,$$

where the sum on the right side refers to the additive group of the model $D^{(I)}$. Moreover, $J_q\omega$ is unique.

4.9(B_q) Any (q-1) bud over K_o has a q-bud extension, i.e. can be represented by a morphism defining a q-bud.

4.10(C_q) Let f be a morphism defining a q-bud over K and g' be a morphism defining a (q-1)-bud over K_o, such that $\varphi_* g' \equiv f$ mod.deg. q.

Then there is a morphism g defining a q-bud over K_o, such that

$$g' \equiv g \quad \text{mod.deg. } q \quad \text{and} \quad \varphi_* g \equiv f \quad \text{mod.deg. } (q+1) .$$

5. Obstructions

The prerequisites from homological algebra are practically reduced to nothing. Still, for some readers, it may be convenient to write down the general formula for the coboundary operator

δ , namely

5.1 $$\delta f(x_1,\dots,x_{n+1}) = f(x_2,\dots,x_{n+1}) + \sum_{1\le i\le n}(-1)^i f(x_1,\dots,x_i+x_{i+1},\dots,x_{n+1}) + (-1)^{n+1} f(x_1,\dots,x_n) .$$

5.2 _The homomorphism obstruction lemma_. _Let_ $f : V \times V \to V$ _and_ $g : W \times W \to W$ _be two morphisms of formal varieties, such that_ $f(x,y) \equiv x + y$ mod. deg. 2 _and_ $g(x,y) \equiv x + y$ mod. deg. 2. _Let_ n _be some integer_ > 1 . _For any morphism_ $u : V \to W$ _such that_

$$u(f(x,y)) \equiv g(u(x),u(y)) \quad \text{mod. deg. } n ,$$

we put

5.3 $$\Delta(u) = \mathrm{dif}_n(u(f(x,y)),\ g(u(x),u(y))) .$$

Then for any morphism $u' : V \to W$, _such that_ $u' \equiv u$ mod. deg. n, _we have_

5.4 $$\Delta(u') = \Delta(u) - \delta h ,$$

where $h = \mathrm{dif}_n(u',u)$ _and_ $\delta h(x,y) = h(x) + h(y) - h(x+y)$.

This lemma is an easy consequence of formula (I.9.10). Indeed, we obtain

$\mathrm{dif}_n(u'(f(x,y)),\ u(f(x,y)) = h(x+y)$, from $h = \mathrm{dif}_n(u',u)$ and $f(x,y) \equiv x + y$ mod. deg. 2 ; also $\mathrm{dif}_n(g(u'(x),\ u'(y)),\ g(u(x),\ u(y)) = h(x) + h(y)$, from $g(x,y) \equiv x + y$ mod. deg. 2 .

5.5 _The bud obstruction lemma_. _For any morphism_ $f : V \times V \to V$ _defining a_ (n-1)-_bud_, _we put_

5.6 $$\Gamma_n f(x,y,z) = \mathrm{dif}_n(f(f(x,y),z),\ f(x,f(y,z))) .$$

Let $f' : V \times V \to V$ _define the same_ (n-1)-_bud_, _with_ $h = \mathrm{dif}_n(f',f)$. _Then_

5.7 $$\Gamma_n f' = \Gamma_n f - \delta h ,$$

where $\delta h(x,y,z) = h(y,z) - h(x+y,z) + h(x,y+z) - h(x,y)$.

Proof. From formula (I.9.10) and the relation $f(x,y) \equiv x + y$ mod. deg. 2, we obtain

$$\mathrm{dif}_n(f'(f'(x,y),z),\ f(f(x,y),z)) = h(x+y,z) + h(x,y) ,$$

$$\mathrm{dif}_n(f'(x,f'(y,z)),\ f(x,f(y,z))) = h(x,y+z) + h(y,z) ,$$

whence formula (5.7) follows.

Now we can discuss two <u>elementary extension problems</u>. In lemma (5.2), let us assume that f <u>and</u> g <u>both define</u> n-<u>buds</u>. Then u defines a (n-1)-bud homomorphism (4.5), which we try to extend to a n-bud homomorphism by adding a correcting term $h \in \mathfrak{P}_n(\mathfrak{X}V,\mathfrak{X}W)$. For this to hold, h has to satisfy

5.8 $$\delta h = \Delta(u) .$$

In lemma (5.5), f defines a (n-1)-bud, which we want to extend to a n-bud by adding a correcting term $h \in \mathfrak{P}_n(\mathfrak{X}V\times\mathfrak{X}V,\mathfrak{X}V)$. For this to hold, h has to satisfy

5.9 $$\delta h = \Gamma_n f .$$

From the coboundary relation $\delta\delta = 0$, we obtain conditions for the solvability of equations (5.8) and (5.9), namely

5.10 $$\delta\Delta(u) = 0 \quad \text{and} \quad \delta\Gamma_n f = 0 , \quad \text{respectively} .$$

Now, conditions (5.10) are always satisfied, even for non necessarily commutative buds, but this fact will not be needed, and we leave the proofs to the reader. Indeed, we need not know the relation $\delta\delta = 0$, except when applied to a 1-cochain, i.e. a $h \in \mathfrak{P}_n(\mathfrak{X}V,\mathfrak{X}W)$, where it is easily checked by direct inspection.

Here we shall be content with determining the first cohomology groups, i.e. the 1-cocycles, or the $h \in \mathfrak{P}_n(\mathfrak{T}V,\mathfrak{T}W)$ such that $\delta h = 0$. For this, we introduce coordinates on V, say $V = D^{(I)}$. Then any $h \in \mathfrak{P}_n(\mathfrak{T}V,\mathfrak{T}W)$ is written uniquely (see I.5.9) as

5.11 $$h(x) = \Sigma_{|\alpha|=n} c_\alpha x^\alpha , \quad \alpha \in \underline{\underline{N}}^{(I)} , \quad c_\alpha \in \mathfrak{T}W .$$

For an $x = (x_i)_{i\in I}$, we write

5.12 $$x = \Sigma_{i\in I} \varepsilon_i(x_i) ,$$

thereby defining the canonical curves $\varepsilon_i : D \to D^{(I)}$. Note that the right side of (5.12) converges for the simple topology, not for the order topology if I is infinite. If h is a 1-cocycle, which means only that h is additive, we obtain from (5.12)

5.13 $$h(x) = \Sigma_{i\in I} c_i x_i^n , \quad c_i \in \mathfrak{T}W .$$

Then we have

5.14 $$\delta h(x,y) = - \Sigma_{i\in I} c_i B_n(x_i,y_i) ,$$

where B_n is the polynomial with coefficients from $\underline{\underline{Z}}$ defined by

5.15 $$B_n(x,y) = (x+y)^n - x^n - y^n = \Sigma_{0<i<n} \binom{n}{i} x^i y^{n-i} .$$

5.16 <u>For any</u> $n \geq 2$, <u>we define</u> $\eta_n \in \underline{\underline{P}}$ as the <u>greatest common divisor of all coefficients of</u> B_n, <u>and we put</u>

5.17 $$B_n = \eta_n C_n .$$

In this way, $C_n(x,y) \in \underline{\underline{Z}}[x,y]$, and the coefficients of C_n are <u>relatively prime</u>. Now we can write (5.14) as

5.18 $$\delta h(x,y) = - \Sigma_{i\in I} C_n(x_i,y_i) \eta_n c_i , \quad c_i \in \mathfrak{T}W ,$$

and the cocycle condition $\delta h = 0$ becomes equivalent to

5.19 <u>for every</u> $i \in I$, $\eta_n c_i = 0$.

Remember that $\mathfrak{T}W$ is always a free K-module.

5.20 If the basic ring K is torsion-free, then $h \in \mathfrak{P}_n(\mathfrak{T}V, \mathfrak{T}W)$, $n \geq 2$ and $\delta h = 0$ imply $h = 0$. This follows immediately from (5.19) and $\eta_n \in \underline{\underline{P}}$, so that equation (5.8) cannot have more than one solution h. We express this fact as follows.

5.21 Lemma. Let the basic ring K be torsion-free. Then the functor $\mathfrak{T}$ is faithful (but, in general, not fully faithful) on the category of formal groups and on the category of n-buds (for any $n \in \underline{\underline{P}}$) over K.

This proves already a part of the induction hypothesis (A_q) (see 4.8), namely the unicity of $J_q\omega$.

6. The 1-dimensional symmetric 2-cocycle lemma

6.1 Lemma. Let n be any integer and A any additive group. Then any symmetric 2-cocycle of degree n in two letters, say $P(x,y) \in A[x,y]$, can be written as

$P(x,y) = c\, C_n(x,y)$, with a unique $c \in A$ (see 5.17).

This lemma means that, if

6.2 $$P(x,y) = \Sigma_{i+j=n}\, a_{i,j}\, x^i y^j, \quad a_{i,j} \in A,$$

satisfies the 2-cocycle condition, namely

6.3 $$P(y,z) - P(x+y,z) + P(x,y+z) - P(x,y) = 0,$$

and the symmetry condition, namely

6.4 $P(x,y) = P(y,x)$, or equivalently $a_{i,j} = a_{j,i}$, then $P = c\, C_n$ for some well defined $c \in A$.

By putting $x = y = 0$ or $y = z = 0$ in (6.3), we obtain

6.5 $$a_{o,n} = a_{n,o} = 0.$$

Then equation (6.3) is expressed by the relations

6.6 $$\binom{i+j}{j} a_{i+j,k} = \binom{j+k}{j} a_{i,j+k} \text{ , for } i+j+k=n \text{ , } i>0, k>0$$

We shall use equation (6.6) only for $i = 1$, when it reduces to

6.7 $$(j+1)\, a_{j+1,k} = \binom{n-1}{j} a_{1,n-1} \text{ , \underline{for} } 0 \leq j \leq n-2 \text{ \underline{and} } j+k=n-1.$$

6.8 Let us first assume that, <u>for some given prime</u> p, $pc = 0$ <u>for any</u> $c \in A$ (i.e. A is a vector space on the prime field $\mathbb{F}_p$) .

6.9 <u>First case</u>: $p|n$ <u>and</u> $n \neq p$. Then we can put $j = p-1$ in (6.7), and, <u>from the elementary relation</u> $\binom{n-1}{p-1} \not\equiv 0$ mod.p, we obtain $a_{1,n-1} = 0$. Further, equation (6.7) gives

$$a_{j+1,k} = 0 \quad \text{for} \quad j+1 \not\equiv 0 \quad \text{mod.p} \text{ ,}$$

so that $P(x,y)$ can be written uniquely as

6.10 $$P(x,y) = Q(x^p, y^p) \text{ ,}$$

where Q is a polynomial of degree n/p. From the additivity of $x \mapsto x^p$ in characteristic p, we see that Q has to be a symmetric 2-cocycle of degree n/p, with coefficients from A .

6.11 <u>Second case</u>: $n = p$. Then a direct inspection shows that $\eta_p = p$ (see 5.16), and that $C_p(x,y) = x^{p-1}y + \ldots + xy^{p-1}$. If we replace P by $P - a_{1,p-1} C_p$, we can assume that $a_{1,p-1} = 0$, whence $(j+1)a_{j+1,p-j-1} = 0$ for $0 \leq j \leq p-2$, by (6.7) .

6.12 <u>Third case</u>: $p \nmid n$. Then $B_n(x,y) = nx^{n-1}y + \ldots + nxy^{n-1}$, and $C_n(x,y) = n'x^{n-1}y + \ldots + n'xy^{n-1}$, with $n' = \eta_n^{-1}n$, whence $n' \not\equiv 0$ mod p. By replacing P by $P - c\, C_n$ for a suitable $c \in A$, we can achieve that $a_{1,n-1} = 0$. Then equation (6.7) gives

6.13 $$a_{i,j} = 0 \quad \text{for} \quad i \not\equiv 0 \quad \text{mod.p} \text{ .}$$

But we have the symmetry condition (6.4), so that $a_{i,j} = a_{j,i}$, and, as $i+j = n$ and $p \nmid n$, we have $p \nmid i$ or $p \nmid j$, so that in any case $a_{i,j} = 0$.

6.14 By applying (6.9), and putting in general $n = n_o p^h$, with $p \nmid n_o$, we see that

6.15
$$\begin{cases} P(x,y) = c\, C_{n_o}(x^{p^h}, y^{p^h}) , & \text{for } n_o > 1 , \\ P(x,y) = c\, C_p(x^{p^{h-1}}, y^{p^{h-1}}) , & \text{for } n_o = 1 , \end{cases}$$

with unique $c \in A$. It remains only to check that (6.15) can be written, <u>in characteristic</u> p, as

6.16
$$P(x,y) = c\, C_n(x,y) .$$

This is easy, but unnecessary, for by taking $A = \mathbb{F}_p$, we have proved that the symmetric 2-cocycles of degree n make up a 1-dimensional vector subspace in $\mathbb{F}_p[x,y]$, and that is enough.

Now let us write

6.17
$$C_n(x,y) = \Sigma_{i+j=n}\, c_{i,j}\, x^i y^j .$$

As the coefficients $c_{i,j}$ are <u>relatively prime</u>, there are integers $\lambda_i \in \mathbb{Z}$ such that

6.18
$$\Sigma_{1 \leq i \leq n-1}\, \lambda_i c_{i,j} = 1 .$$

In the statement of lemma (6.1), let us put

6.19
$$c = \Sigma_{1 \leq i \leq n-1}\, \lambda_i a_{i,j} ,$$

where the $a_{i,j}$ are as in (6.2). We shall presently prove that $P' = P - c\, C_n = 0$. Indeed we can assume that the additive group A is generated by the coefficients $a'_{i,j}$ of P'. By reduction modulo any prime p, we know that

6.20
$$P' \equiv c_p\, C_n \ \text{mod.}p , \text{ for some } c_p \in A,$$

and equations (6.18), (6.19) insure that $c_p \equiv 0 \mod p$. But this means that the finitely generated group A is divisible by p for any prime p, which, by the direct decomposition theorem, means that $A = 0$, i.e. the lemma (6.1) is proved.

7. Curvilinear group laws and buds

7.1 In any model $D^{(I)}$, we have the canonical curves $\varepsilon_i : D \to D^{(I)}$ (see 5.12), such that

$$x = (x_i)_{i\in I} = \Sigma_{i\in I}\, \varepsilon_i(x_i) ,$$

and the canonical projections $p_i : D^{(I)} \to D$, defined by

$$p_i(x) = x_i .$$

7.2 Proposition. Let G be a formal group and $(\gamma_i)_{i\in I}$ a basic set of curves in G (see I.10.18). Then the morphism $f : D^{(I)} \to G$, defined by

7.3 $$f = \Sigma_{i\in I}\, \gamma_i \circ p_i ,$$

where the right side converges in the group $\mathfrak{M}(D^{(I)}, G)$ for the simple topology, is an isomorphism, i.e. a coordinate system on G (see I.4.1). This coordinate system will be called the curvilinear coordinate associated with the basic set $(\gamma_i)_{i\in I}$.

Indeed, according to the isomorphism theorem (I.8.1), it suffices to prove that $\mathfrak{T}f$ is an isomorphism. But, from the definition (7.3), it follows that $f \circ \varepsilon_i = \gamma_i$ for any $i \in I$. Therefore $\mathfrak{T}f \cdot \mathfrak{T}\varepsilon_i = \mathfrak{T}\gamma_i$, which means that $\mathfrak{T}f$ maps the canonical basis $(\mathfrak{T}\varepsilon_i)_{i\in I}$ of $\mathfrak{T}D^{(I)}$ onto the basis $(\mathfrak{T}\gamma_i)_{i\in I}$ of $\mathfrak{T}G$.

7.4 Definition. We say that a group law $\mu : D^{(I)} \times D^{(I)} \to D^{(I)}$ is curvilinear iff

$$\Sigma^{\mu}_{i\in I}\ \varepsilon_i \circ p_i = \mathrm{Id} \quad (D^{(I)} \to D^{(I)}) ,$$

<u>where</u> Σ^{μ} <u>means the sum in the group</u> $\mathfrak{M}(D^{(I)}, D^{(I)})$, <u>defined by the group law</u> μ (and not only for the additive group law of $D^{(I)}$, as in 2.21).

7.5 It follows from proposition (7.2) that, <u>for any group law</u> $\mu : D^{(I)} \times D^{(I)} \to D^{(I)}$, <u>there is a unique isomorphism</u> $f : D^{(I)} \to D^{(I)}$, <u>such that</u> $f \circ \varepsilon_i = \varepsilon_i$ <u>for any</u> $i \in I$ (i.e. f leaves the canonical curves of the model invariant), <u>and that</u> $f^{-1}(\mu(f(x), f(y)))$ <u>is a curvilinear group law</u>.

7.6 The definition of curvilinear coordinates holds for any n-bud on a model. The sum $\Sigma\ \varepsilon_i \circ p_i$ is not defined, but its n-jet is defined, and we assume that

$$\Sigma^{\mu}_{i\in I}\ J_n(\varepsilon_i) \circ J_n(p_i) = J_n\ \mathrm{Id} .$$

7.7 <u>Criterion for curvilinear group laws and buds</u>. <u>Let</u> $\mu : D^{(I)} \times D^{(I)} \to D^{(I)}$ <u>be a group law</u> (<u>resp. define a</u> n-<u>bud</u>, <u>for some</u> $n \geq 2$). <u>Explicitely</u>, <u>let</u>

$$\mu(x,y) = \Sigma_{\alpha,\beta\in\underline{\underline{N}}^{(I)}}\ c_{\alpha,\beta}\ x^{\alpha}y^{\beta} ,$$

<u>with</u> $c_{\alpha,\beta} \in K^{(I)}$: <u>see</u> (I.5.2) <u>and</u> (I.7.2). <u>Then</u> μ <u>is curvilinear iff</u>

7.8 $$\alpha,\beta \in \underline{\underline{N}}^{(I)} ,\ |\alpha|+|\beta| \geq 2 \ \underline{\text{and}}\ \ \mathrm{supp}(\alpha) \cap \mathrm{supp}(\beta) = \emptyset$$
$$\underline{\text{implies}}\quad c_{\alpha,\beta} = 0$$

(<u>for an</u> n-bud, <u>condition</u> (7.8) <u>is only assumed to hold if</u> $|\alpha|+|\beta| \leq n$) .

Indeed, let $x = (x_i)_{i\in I}$ and $y = (y_i)_{i\in I}$ be two elements of $A^{(I)}$, for some $A \in \underline{\underline{\mathrm{nil}}}(K)$, such that $\mathrm{supp}(x) \cap \mathrm{supp}(y) = \emptyset$. This means that, for any $i \in I$, we have $x_i = 0$ or $y_i = 0$, so that, putting $z_i = x_i + y_i$, we have

$z_i = y_i$ or $z_i = x_i$. Then we have

7.9 $$z = x + y ,$$

<u>either for the additive group law</u> (2.21), <u>or for the curvilinear group law</u> μ . This is expressed by condition (7.8), and, for a n-bud, we have to assume that $|\alpha|+|\beta| \leq n$. Conversely, condition (7.9) implies that, for any $A \in \underline{\underline{nil}}(K)$ and any $x \in A^{(I)}$, we

$$x = \Sigma^{\mu}_{i \in I} \; \varepsilon_i(x_i)$$

for the group law (resp. n-bud) μ , which, accordingly is curvilinear.

7.10 <u>Lemma</u>. <u>Let</u> $\omega : D^{(I)} \to D^{(I)}$ <u>be an isomorphism and</u> $\mu(x,y) = \omega^{-1}(\omega(x) + \omega(y))$, <u>using the additive group on</u> $D^{(I)}$.

<u>Then</u> μ <u>is a curvilinear group law</u> (<u>resp. defines a curvilinear</u> n-<u>bud</u>) <u>iff</u> $\omega = \Sigma_{i \in I} \; \omega \circ \varepsilon_i \circ p_i$ (<u>resp</u>. $\omega \equiv \Sigma_{i \in I} \; \omega \circ \varepsilon_i \circ p_i$ mod.deg. (n+1)).

This is proved as the criterion (7.7), by applying the relation $\omega(\mu(x,y)) = \omega(x) + \omega(y)$ to $x,y \in A^{(I)}$, such that $\mathrm{supp}(x) \cap \mathrm{supp}(y) = \emptyset$.

7.11 <u>The curvilinear</u> 2-<u>cocycle lemma</u>. <u>Let</u> $P(x,y) = \Sigma_{\alpha,\beta \in \underline{\underline{N}}^{(I)}, |\alpha|+|\beta|=n} \; c_{\alpha,\beta} \, x^{\alpha} y^{\beta}$, $c_{\alpha,\beta} \in K^{(I)}$ <u>be a</u> 2-<u>cocycle</u>, <u>i.e. verify</u>

$$P(y,z) - P(x+y,z) + P(x,y+z) - P(x,y) = 0 .$$

<u>Assume that</u> $\mathrm{supp}(\alpha) \cap \mathrm{supp}(\beta)$ <u>implies</u> $c_{\alpha,\beta} = 0$. <u>Then we have</u>

$$P(x,y) = \Sigma_{i \in I} \; c_i P_i(x_i,y_i) \quad , \quad c_i \in K^{(I)} ,$$

<u>where the</u> P_i <u>are</u> 2-<u>cocycles of degree</u> n <u>in two letters</u>.

Proof. The cocycle condition for P implies $c_{\alpha,\beta} = 0$ for $|\alpha| = 0$ or $|\beta| = 0$, as in (6.5), and is written (as in

6.6)

7.12 $$\binom{\alpha+\beta}{\beta} c_{\alpha+\beta,\gamma} = \binom{\beta+\gamma}{\gamma} c_{\alpha,\beta+\gamma}\ , \quad \text{for } |\alpha+\beta+\gamma| = n, \quad \alpha \neq 0, \ \gamma \neq 0 .$$

It is clear from formula 7.12 that, in order to check the cocycle condition on the $c_{\alpha,\beta}$, we need only consider separately the $c_{\alpha,\beta}$ with prescribed value $\delta \in \underline{\underline{N}}^{(I)}$ of $\alpha + \beta$. The lemma means that $c_{\alpha,\beta} = 0$ unless there is $i \in I$ such that $\alpha_i + \beta_i = n$. Assume that this is not the case. Then we can write I as $\{i_o\} \cup I'$, $i_o \notin I'$, and write $\alpha \in \underline{\underline{N}}^{(I)}$ as a pair (α_o,α'), with $\alpha_o \in \underline{\underline{N}}$ and $\alpha' \in \underline{\underline{N}}^{(I')}$. With these notations, we shall presently prove that

7.13 $$c_{(\alpha_o,\alpha'),(\beta_o,\beta')} = 0 \ \text{ if } \ \alpha_o + \beta_o > 0,\ \alpha' + \beta' \neq 0 ,$$
$$(\alpha_o,\alpha') \neq (0,0),\ (\beta_o,\beta') \neq (0,0) .$$

Let us assume that $\alpha_o \neq 0$. Then a first application of (7.12), where we replace α,β,γ respectively by $(\alpha_o,0)$, $(0,\alpha')$ and (β_o,β'), gives

7.14 $$c_{(\alpha_o,\alpha'),(\beta_o,\beta')} = \binom{\alpha'+\beta'}{\beta'} c_{(\alpha_o,0),(\beta_o,\alpha'+\beta')} .$$

A second application of (7.12), with α,β,γ replaced respectively by $(\alpha_o,0)$, $(\beta_o,0)$, $(0,\alpha'+\beta')$, gives

7.15 $$\binom{\alpha_o+\beta_o}{\beta_o} c_{(\alpha_o+\beta_o,0),(0,\alpha'+\beta')} = c_{(\alpha_o,0),(\beta_o,\alpha'+\beta')} .$$

As, by hypothesis $c_{(\alpha_o+\beta_o,0),(0,\alpha'+\beta')} = 0$, equations (7.14) and (7.15) imply

$$c_{(\alpha_o,\alpha'),(\beta_o,\beta')} = 0 ,$$

under the assumptions $\alpha_o + \beta_o > 0$, $\alpha' + \beta' \neq 0$. If $\alpha_o = 0$, then one application of (7.12) is sufficient: namely, we replace α,β,γ respectively by $(0,\alpha')$, $(0,\beta')$, $(\beta_o,0)$.

8. End of the proofs of the theorems

8.1 Lemma. *Let* I *be any set*, K *any basic ring*, n *any integer* ≥ 2 *and* $f, f' : D_K^{(I)} \times D_K^{(I)} \to D_K^{(I)}$ *be two morphisms. Assume that* f *defines a curvilinear* n*-bud, and that* $J_{n-1}f = J_{n-1}f'$. *Then* f' *defines a curvilinear* n*-bud iff there are elements* $c_i \in K^{(I)}$, $i \in I$ *(necessarily unique), such that*

8.2 $$f'(x,y) \equiv f(x,y) + \Sigma_{i \in I} \; c_i C_n(x_i, y_i) \quad \text{mod.deg. } (n+1) .$$

Proof. If f' defines a n-bud, then $\mathrm{dif}_n(f',f)$ is a 2-cocycle P, according to lemma (5.5). Moreover, if f and f' define curvilinear n-buds, then the hypothesis of lemma (7.11) holds for P, according to the criterion (7.7). But, as P is symmetric, we can apply lemma (6.1) to every component P_i in the statement of lemma (7.11): this gives formula (8.2).

Conversely, if f' verifies (8.2) for some c_i , then f' defines a n-bud, according to (5.5), and this n-bud is curvilinear, according to (7.7).

8.3 Proof that (A_{n-1}) implies (A_n); see (4.8). In proving (A_n), we apply proposition (7.2), which allows us to assume that $f : D^{(I)} \times D^{(I)} \to D^{(I)}$ defines a curvilinear n-bud over the $\underline{\mathbb{Q}}$-algebra K_1 . Then we apply (A_{n-1}), and take the morphism $\omega : D^{(I)} \to D^{(I)}$ with vanishing coefficients in degrees $\geq n$. In this way the group law f' , defined by

8.4 $$\omega(f'(x,y)) = \omega(x) + \omega(y)$$

is curvilinear (see 7.10).

Now we define $\Delta\omega \in \varphi_n(K^{(I)} \times K^{(I)}, K^{(I)})$ by

8.5 $$\Delta\omega(x,y) \equiv \omega(f(x,y)) - \omega(x) - \omega(y) \quad \text{mod.deg. } (n+1).$$

We wish to replace ω by some ω', defined by

8.6 $$\omega'(x) = \omega(x) + \Sigma_{i\in I}\ x_i^n\ c_i' \ , \quad c_i' \in K^{(I)} ,$$

in such a way that $\Delta\omega' = 0$. By lemma (5.2) and the definition of the polynomials C_n, this is equivalent to

8.7 $$\Sigma_{i\in I} - c_i'\ \eta_n\ C_n(x_i,y_i) = \Delta\omega(x,y) \ .$$

From (8.4), we can write $\Delta\omega$ as

$$\Delta\omega(x,y) = \mathrm{dif}_n(\omega(f(x,y)),\ \omega(f'(x,y))) ,$$

and, as $\omega(x) \equiv x$ mod.deg. 2, we have also (see I.9.10)

$$\Delta\omega(x,y) = \mathrm{dif}_n\ (f,f') \ .$$

We apply lemma (8.1) to f and f', and we see that the c_i' are related to the c_i in (8.2) by

$$c_i' = \eta_n^{-1}\ c_i \quad , \quad i \in I \ .$$

8.8 Proof of the $\underline{\underline{Q}}$-theorem. We have just proved (A_q) for all q (see 4.8). This implies the existence of the isomorphism (3.5) for any formal group over a $\underline{\underline{Q}}$-algebra. Therefore the map $\mathfrak{T}$ from $\mathrm{Hom}(G,G')$ to $\mathfrak{L}_K(\mathfrak{T}G,\mathfrak{T}G')$, the set of K-linear maps, is surjective for two formal groups G,G'. But we know that it is injective (see 5.21).

8.9 Proof of (B_n); see (4.9). Let $f : D_{K_o}^{(I)} \times D_{K_o}^{(I)} \to D_{K_o}^{(I)}$ define a (n-1)-bud over the torsion-free ring K_o. We can assume that this bud is curvilinear, and that f has vanishing terms in degrees $\geqslant n$. So we want to find $h \in \mathfrak{P}_n(K_o^{(I)} \times K_o^{(I)},\ K_o^{(I)})$, in such a way that f', given by

$$f'(x,y) = f(x,y) + h(x,y)$$

defines a n-bud.

Let us put

8.10 $$h(x,y) = \Sigma_{\alpha,\beta\in\underline{\underline{N}}^{(I)},|\alpha+\beta|=n}\ c_{\alpha,\beta}\ x^\alpha y^\beta \ ,$$

and, for every $\gamma \in \underline{\underline{N}}^{(I)}$, $|\gamma| = n$,

$$h_\gamma(x,y) = \Sigma_{\alpha+\beta=\gamma} \; c_{\alpha,\beta} \, x^\alpha y^\beta \; .$$

Then we know from lemma (5.5) that h has to satisfy an equation of the form

8.11 $$\delta h = \Gamma_n f \; ,$$

which, on closer inspection, splits as a set of equations

8.12 $$\delta h_\gamma = \Gamma_{n,\gamma} f \; ,$$

where $\gamma \in \underline{\underline{N}}^{(I)}$, $|\gamma| = n$, and $\Gamma_{n,\gamma} f$ denotes the sum of terms of total multidegree γ in $\Gamma_n f$.

Now, we consider K_o as a subring of $K_1 = \underline{\underline{Q}} \otimes K_o$, and we apply (A_{n-1}) to the (n-1)-bud over K_1 defined by f. We see that f is isomorphic to $x + y$ mod.deg. n and over K_1 : this means that the equations (8.12) have solutions with coefficients from K_1. But, as there is only a finite number of elements from K_1 appearing in any given h_γ , we can write, for $\gamma \in \underline{\underline{N}}^{(I)}$, $|\gamma| = n$.

8.13 $$h_\gamma = n_\gamma^{-1} \, h'_\gamma \; ,$$

with $h'_\gamma \in \mathfrak{P}_n(K_o^{(I)} \times K_o^{(I)}, K_o^{(I)}$ and $n_\gamma \in \underline{\underline{P}}$.

What we want to show is that, eventually after correcting the h_γ , they have actually coefficients from K_o . Denoting by a bar the reduction modulo n_γ , we see from (8.13) that h_γ has coefficients from K_o iff $\overline{h'_\gamma} = 0$. But (8.12) implies that

8.14 $$\delta \overline{h'}_\gamma = 0 \; .$$

We can assume that $f' = f + h$ defines a curvilinear n-bud. Then we can apply the lemmas (6.1) and (7.11) to $\overline{h'_\gamma}$, over the ring $\overline{K}_o = K_o/n_\gamma K_o$. We see that $\overline{h'_\gamma} = 0$, unless there is some $i \in I$ such that $\gamma_i = n$; in this last case, there is a $c_i \in K_o$ such that $\overline{h'_\gamma} = \bar{c}_i\, C_n(x_i, y_i)$, and we write n_i instead of n_γ. In this way, we replace $h(x,y)$ by $h(x,y) - \Sigma_{i\in I}\, n_i^{-1}\, c_i\, C_n(x_i, y_i)$, and we obtain what we wanted: a solution to (8.11) with coefficients from K_o.

8.15 By proving (B_n) for all n, we have proved the extension theorem (4.6) for a torsion-free basic ring.

8.16 Proof of (C_n). In the statement (4.10), where we put n instead of q, we can assume that g' defines a curvilinear n-bud over K_o, according to (B_n), and that f defines a curvilinear n-bud over K. Then, putting $f' = \varphi_* g'$, we apply lemma (8.1), which shows that $\mathrm{dif}_n(f', f)$ can be lifted over K_o, thereby obtaining the required g over K_o.

8.17 The lift theorem (3.8) is a consequence of "(C_n) for all n" when the ring K_o in (3.8) is assumed torsion-free. Indeed, we have proved the lift theorem for buds.

But, if $\varphi : K_o \to K$ is a surjective homomorphism, we can take $\varphi' : K'_o \to K_o$, a surjective ring homomorphism with torsion free K'_o. We can apply the lift theorem to $\varphi_1 = \varphi \circ \varphi'$ and afterwards apply φ'_* to the lifted group-law over K'_o. This proves the lift theorem.

Finally, the extension theorem (4.6) is a consequence of the lift theorem for buds, and of (8.15).

9. A digression concerning non-commutative groups

9.1 Here we discuss "non necessarily commutative" formal groups, so that our general convention (2.1) is cancelled, just for a while.

The method we used in the present chapter, which can be summarized by the formula "$J_\infty = \varprojlim J_n$", leads generally to non-trivial obstructions, so that the lift theorem (3.8) and the extension theorem (4.6) are no more true when the commutativity assumption is dropped.

9.2 But the $\underline{Q}$-theorem (3.1) holds mutatis mutandis, the change beeing the introduction of a Lie algebra structure on the tangent space of a formal group.

9.3 Definition. Let G be a formal group over K. Then the Lie algebra of G is its tangent space $\mathfrak{T}G$, together with the hessian of the commutator morphism (see I.9.2), $\mathfrak{T}G \times \mathfrak{T}G \to \mathfrak{T}G$, written $x,y \mapsto [x,y]$.

Let us assume that the group morphism f of G is defined on a formal module L^+, so that

9.4 $$f(x,y) \equiv x + y + a(x,y) \quad \text{mod. deg. } 3,$$

where a is the component of total degree 2. As a consequence of the relations $f(x,0) = f(0,x) = x$, we have $a(x,0) = a(0,x) = 0$, so that a is a bilinear morphism, i.e. a is homogeneous of bidegree (1,1).

So f corresponds to the multiplicative word xy. By definition, the commutator is the word $x^{-1}y^{-1}xy$, which we write as $(yx)^{-1}(xy)$. The morphism corresponding to the word yx is $f(y,x)$, and $f(y,x) \equiv f(x,y)$ mod. deg. 2. Therefore the hessian of the commutator morphism is

9.5 $$\mathrm{dif}_2(f(x,y),\ f(y,x)) = a(x,y) - a(y,x) = [x,y] ,$$

because the difference in degree 2 may be computed by (9.3) or by using the morphism corresponding to the word $y^{-1}x$ (see I. 9.10).

Clearly, the Lie bracket $[x,y]$ is <u>bilinear</u> and <u>alternate</u>. There are many ways to prove that it verifies the <u>Jacobi identity</u>. One of them is to use an identity of Philip Hall: when writing (x,y) for the commutator $x^{-1}y^{-1}xy$ and x^y for $y^{-1}xy$, one has (in a multiplicative group)

9.6 $$(x^y,(y,z))(y^z,(z,x))(z^x,(x,y)) = 1 .$$

As $x^y = x(x,y)$, the corresponding word morphism is $x + [x,y]$ mod. deg. 3. The word morphism corresponding to the left side of (9.6) has to vanish, and it is

$$[x,[y,z]] + [y,[z,x]] + [z,[x,y]] \quad \text{mod. deg. } 4 ,$$

which proves Jacobis identity.

Another method is to consider the 2-bud defined by $x + y + a(x,y)$. Then Γ_3, computed by (5.6), is

9.7 $$a(a(x,y),z) - a(x,a(y,z)) = b(x,y,z)$$

It vanishes iff a is associative; that is the condition for $x + y + a(x,y)$ to be a group law. But the existence of a 3-bud extending the 2-bud $x + y + a(x,y)$ is equivalent to that of a morphism $h \in \mathfrak{P}_3(L\times L,L)$, such that

9.8 $$\delta h(x,y,z) = b(x,y,z) \qquad \text{(see 5.5)} .$$

The only bihomogeneous components of h that matter are $h_{2,1}(x,y)$ and $h_{1,2}(x,y)$, and, from the coboundary formula, it is easy to deduce that the <u>antisymmetrized</u> morphism of b has to vanish. Writing that down, one obtains a sum of 12 terms, which gives Jacobi's indentity when introducing the "bracket" $[x,y] = a(x,y) - a(y,x)$.

9.9 <u>The $\underline{\underline{Q}}$ theorem for non commutative formal groups</u>. <u>Let</u> K <u>be a</u> $\underline{\underline{Q}}$-<u>algebra</u>, G <u>and</u> G' <u>two formal groups over</u> K <u>and</u>

$u : \mathfrak{L}G \to \mathfrak{L}G'$ a Lie algebra homomorphism. Then there is one and only one formal group homomorphism $f : G \to G'$ such that $\mathfrak{L}f = u$.

That can be proved from lemma (5.2), by showing that, over the $\underline{\underline{Q}}$-algebra K, any 2-cocycle of degree ≥ 3 is a 2-coboundary.

9.10 In order to prove that the category of formal groups over a $\underline{\underline{Q}}$-algebra K is equivalent to that of Lie algebras over K (which are free qua K-modules), it suffices to prove the existence of a formal group with given Lie algebra. Such a group is defined by Hausdorff's formula. For the proofs, see [10] or [3].

9.11 Theorem. One-dimensional formal groups over a basic ring K are commutative, except when K contains a non-zero nilpotent element of finite additive order.

In the exceptional case, there is some $a \in K$, such that $a^2 = 0$ and $pa = 0$ (p prime); then $x + y + axy^p$ is a non-commutative group law.

For a proof of (9.11), see [14] and [6].

CHAPTER III

THE GENERAL EQUIVALENCE OF CATEGORIES

1. Definition of W^+ and $\hat{W}^+$ from $\mathfrak{C}(\underline{\underline{G}}_m)$

1.1 We have already introduced the additive group law $\underline{\underline{G}}_a$ and the multiplicative group law $\underline{\underline{G}}_m$ (see II.2.12). Both are defined over $\underline{\underline{Z}}$, therefore over any basic ring K (see I.11.17 and II.2.11).

1.2 For any basic ring K, we have two topological groups, $\mathfrak{C}_K(\underline{\underline{G}}_a)$ and $\mathfrak{C}_K(\underline{\underline{G}}_m)$, with the same underlying uniform space, namely the set of morphisms $\mathfrak{M}(D_K, D_K)$ with its (simple or order) topology (once the topological group structure is given - no matter which one - there is no difference between a topological space and a uniform space; remember that groups are commutative!). This holds more generally for any one-dimensional group law G over K. According to the morphism lemma (I.3.2), any curve $\gamma \in \mathfrak{C}(G)$ has a well defined <u>expression</u> as a formal series

$$\Sigma_{n\in\underline{\underline{P}}} \; a_n t^n \quad , \quad a_n \in K .$$

1.3 We shall denote by $\gamma_{\underline{\underline{a}}}$ and $\gamma_{\underline{\underline{m}}}$ the <u>identity morphism</u> (expressed by t), <u>qua element of</u> $\mathfrak{C}(\underline{\underline{G}}_a)$ <u>and</u> $\mathfrak{C}(\underline{\underline{G}}_m)$ <u>respectively</u>, with the following convention: any occurence of $\gamma_{\underline{\underline{a}}}$ (resp. $\gamma_{\underline{\underline{m}}}$) in a formula concerning some group $\mathfrak{C}(G)$ shall mean that $G = \underline{\underline{G}}_a$ (resp. $= \underline{\underline{G}}_m$).

For instance, the formula

1.4 $$\gamma = \Sigma_{n\in\underline{\underline{P}}} \; V_n[a_n]\cdot \gamma_{\underline{\underline{a}}} \quad , \quad a_n \in K ,$$

means only that $\gamma \in \mathfrak{C}_K(\underline{\underline{G}}_a)$, and that the expression of γ is given by

1.5 $$\gamma(t) = \Sigma_{n\in\underline{\underline{P}}} \; a_n t^n .$$

The formula in $\mathfrak{C}(\underline{\underline{G}}_m)$ analogous to (1.4), namely

1.6 $$\gamma = \Sigma_{n\in\underline{\underline{P}}} \; V_n[x_n]\cdot \gamma_{\underline{\underline{m}}} ,$$

means that the expression (1.5) of γ is given by the following equality of formal series:

1.7 $$1 - \Sigma_{n\in\underline{\underline{P}}} \; a_n t^n = \Pi_{n\in\underline{\underline{P}}}(1-x_n t^n) .$$

From (1.7), we obtain

1.8
$$\begin{cases} a_1 = x_1 & , \quad x_1 = a_1 \\ a_2 = x_2 & , \quad x_2 = a_2 \\ a_3 = x_3 - x_1x_2 & , \quad x_3 = a_3 + a_1a_2 \quad , \\ a_4 = x_4 - x_1x_3 & , \quad x_4 = a_4 + a_1a_3 + a_1^2a_2 \\ \text{etc.} & \end{cases}$$

1.9 A direct inspection shows what could be derived from more general considerations (see later 6.1). Namely each one of the sequences $a = (a_n)_{n \in \underline{\underline{P}}}$ and $x = (x_n)_{n \in \underline{\underline{P}}}$ can be computed from the other. More precisely, if we give to a_n and x_n the weight n, then a_n (resp. x_n) is expressed by an isobaric polynomial of weight n in the x_i (resp. in the a_i), with coefficient from $\underline{\underline{Z}}$, the coefficient of x_n (resp. a_n) being 1.

Now let us take two curves $\gamma, \gamma' \in \mathfrak{C}_K(\underline{\underline{G}}_m)$ in the form (1.6), namely

$$\gamma = \Sigma_{n \in \underline{\underline{P}}} V_n[x_n] \cdot \gamma_{\underline{\underline{m}}} \quad , \quad \gamma' = \Sigma_{n \in \underline{\underline{P}}} V_n[y_n] \cdot \gamma_{\underline{\underline{m}}} \; , \; x_n, y_n \in K \; .$$

We write their sum $\gamma + \gamma'$ in the same form, namely

$$\gamma + \gamma' = \Sigma_{n \in \underline{\underline{P}}} V_n[z_n] \cdot \gamma_{\underline{\underline{m}}} \; .$$

The relation between the sequences $x = (x_n)_{n \in \underline{\underline{P}}}$, $y = (y_n)_{n \in \underline{\underline{P}}}$ and $z = (z_n)_{n \in \underline{\underline{P}}}$ is given by

1.10
$$\Sigma_{n \in \underline{\underline{P}}} V_n([x_n] + [y_n]) \cdot \gamma_{\underline{\underline{m}}} = \Sigma_{n \in \underline{\underline{P}}} V_n[z_n] \cdot \gamma_{\underline{\underline{m}}} \; ,$$

or, alternatively, by the following equality of formal series

1.11
$$\Pi_{n \in \underline{\underline{P}}} (1-x_nt^n)(1-y_nt^n) = \Pi_{n \in \underline{\underline{P}}} (1-z_nt^n) \; .$$

From (1.10) and (1.8) we obtain

1.12
$$\begin{cases} z_1 = x_1 + y_1 \\ z_2 = x_2 + y_2 - x_1y_1 \\ z_3 = x_3 + y_3 - x_1y_1(x_1+y_1) \\ z_4 = x_4 + y_4 - x_2y_2 + x_1y_1(x_2+y_2 - (x_1+y_1)^2) \text{ , etc.} \end{cases}$$

1.13 A direct argument shows that, <u>if</u> x_n <u>and</u> y_n <u>are given the weight</u> n, <u>then</u> z_n <u>is expressed by an isobaric polynomial of weight</u> n <u>in the</u> x_i <u>and</u> y_i , <u>with coefficients from</u> $\underline{\underline{Z}}$, <u>the coefficients of</u> x_n <u>and</u> y_n <u>being both</u> 1.

1.14 <u>Definition</u>. We shall denote by $W^+(K)$, and call the <u>additive group of general Witt vectors</u> with coefficients from K, the set $K^{\underline{\underline{P}}}$ (of all unrestricted sequences in K), endowed with the group operation defined by (1.11).

We can view W^+ as a functor in groups, namely $K \mapsto W^+(K)$. <u>The group</u> $W^+(K)$ <u>is isomorphic to the group</u> $\mathbb{C}_K(\underline{\underline{G}}_m)$, over $\underline{\underline{Z}}$: see (1.7), (1.11).

1.15 For any $A \in \underline{\underline{nil}}(K)$, we have a group structure on the unrestricted power $A^{\underline{\underline{P}}}$, given by the formulas (1.12). Then the restricted power $A^{(\underline{\underline{P}})}$ becomes a <u>subgroup</u> of the former. Indeed the condition that $x \in A^{\underline{\underline{P}}}$ has a finite support can be expressed by saying that all monomials in the x_n of sufficiently large weight vanish, each x_n having the weight n: the subgroup property follows from (1.13). In other words, we have a group law on the model $D^{(\underline{\underline{P}})}$, defined over $\underline{\underline{Z}}$.

1.16 <u>Definitions</u>. <u>We denote by</u> α_W <u>the group law</u> $D^{(\underline{\underline{P}})} \times D^{(\underline{\underline{P}})} \to D^{(\underline{\underline{P}})}$ <u>given by</u> (1.10). <u>The formal group on the model</u> $D^{(\underline{\underline{P}})}$, <u>defined by</u> α_W, <u>will be denoted by</u> $\hat{W}^+$ <u>and called the formal (additive) group of general Witt vectors</u>.

1 17 There is another way to define $\hat{W}^{+}$. Namely we can first take the group law $D^{(\underline{\underline{P}})} \times D^{(\underline{\underline{P}})} \to D^{\underline{\underline{P}}}$, defined quite explicitely by

$$c_n = a_n + b_n - \Sigma_{1\leq i\leq n-1}\, a_i b_{n-i} \ ,$$

and then replace it by the corresponding curvilinear group, as in (II.7.5).

2. The formal group homomorphism $\underline{\underline{w}} : \hat{W}^{+} \to D_{+}^{(\underline{\underline{P}})}$

2.1 The *derivative* of a formal series $f(t) = \Sigma_{n\in\underline{\underline{N}}}\, a_n t^n$ is, by definition, the formal series $\Sigma_{n\in\underline{\underline{P}}}\, n\, a_n t^{n-1} = f'(t)$. The classical rules hold:

$$(f+g)' = f' + g' \quad \text{and} \quad (fg)' = f'g + fg' \ ;$$

from this one obtains, for invertible f and g ,

$$(fg)'/fg = f'/f + g'/g \ ,$$

i.e. the property of the *logarithmic derivative*.

2.2 For any curve $\gamma \in \mathfrak{C}(\underline{\underline{G}}_m)$, we define the curve $\partial\gamma \in \mathfrak{C}(\underline{\underline{G}}_a)$ by giving its expression:

2.3 $$\partial\gamma(t) = t\ \gamma'(t)(1-\gamma(t))^{-1} \ .$$

In other words, $\partial\gamma$ is the *logarithmic derivative* of $1-\gamma$, *multiplied by* $-t$. Using the expansion $(1-x)^{-1} = \Sigma_{i\in\underline{\underline{N}}}\, x^i$, we can rewrite (2.3) as

2.4 $$\partial\gamma(t) = (\Sigma_{n\in\underline{\underline{P}}}\, n\, a_n t^n)((\Sigma_{i\in\underline{\underline{N}}}\, (\Sigma_{n\in\underline{\underline{P}}}\, a_n t^n)^i) \ ,$$

when $\gamma(t) = \Sigma\, a_n t^n$. Formula (2.4) shows that the coefficients of the formal series $\partial\gamma$ are isobaric polynomials in the a_i , with coefficients from $\underline{\underline{Z}}$ or rather from $\underline{\underline{Z}}\ 1_K$.

2.5 *For any basic ring*, $\partial : \mathfrak{C}_K(\underline{\underline{G}}_m) \to \mathfrak{C}_K(\underline{\underline{G}}_a)$, *is a homomorphism of topological groups*.

Indeed, the additivity of ∂ comes from the property of the logarithmic derivative, and its continuity can be expressed by stating that, for any $n \in \mathbf{P}$, $J_n \, \partial\gamma$ is a function of $J_n \gamma$ (because we multiplied the derivative by the variable).

Let us compute $\partial\gamma$ for $\gamma = V_n[x_n] \cdot \gamma_{\mathbf{m}}$, i.e. $\gamma(t) = x_n t^n$. Then (2.4) reduces to

2.6 $$\partial(V_n[x_n] \cdot \gamma_{\mathbf{m}})(t) = \Sigma_{i \in \mathbf{P}} \, n \, x_n^i t^{ni} .$$

Now, if $\gamma = \Sigma_{n \in \mathbf{P}} V_n[x_n] \cdot \gamma_{\mathbf{m}}$, we obtain, from (2.5) and (2.6)

2.7 $$\partial\gamma(t) = \Sigma_{n \in \mathbf{P}} (\Sigma_{d|n} \, d \, x_d^{n/d}) t^n ,$$

where "$\Sigma_{d|n}$" means a sum ranging over the set of divisors of n in $\mathbf{P}$.

According to (1.9) and (1.15), the additivity of ∂ can be expressed as follows.

2.8 *There is a formal group homomorphism* $\mathbf{w} : \hat{W}^+ \to D_+^{(\mathbf{P})}$, where $D_+^{(\mathbf{P})}$ *denotes the ordinary additive formal group on* $D^{(\mathbf{P})}$, *defined explicitely by its components*

2.9 $$w_n(x) = \Sigma_{d|n} \, d \, x_d^{n/d} \quad , \quad x = (x_n)_{n \in \mathbf{P}} .$$

The simple convergence of w_n towards 0 is insured because $w_n(x)$ is an *isobaric polynomial of weight* n (each x_d having the weight d).

2.10 Let us stress that $\mathbf{w}$ (2.8) is a homomorphism, not an isomorphism in general. More precisely, $\mathbf{w}$ *is an isomorphism iff the basic ring* K *is a* $\mathbf{Q}$*-algebra*, because $\mathfrak{L}\mathbf{w}$ is the linear endomorphism of $K^{(\mathbf{P})}$ multiplying by n the n-th component of each vector (see I.8.1).

2.11 Let $x = (x_n)_{n \in \mathbf{P}}$, $y = (y_n)_{n \in \mathbf{P}}$, $z = (z_n)_{n \in \mathbf{P}}$ be related as in (1.10). Then, for any $n \in \underline{\underline{P}}$,

2.12 $$\Sigma_{d|n}\, d(x_d^{n/d} + y_d^{n/d}) = \Sigma_{d|n}\, dz_d^{n/d}.$$

2.13 These last formulas show, by an easy induction, that, for any $n \in \underline{\underline{P}}$, z_n <u>can be written as a polynomial in the</u> x_d , y_d <u>where</u> $d|n$, and coefficients from $\underline{\underline{Q}}$. But we know already, from (1.11), that z_n can be written as a polynomial in the x_i , y_i where $i \leq n$ and coefficients from $\underline{\underline{Z}}$. Combining these informations, we obtain the following proposition.

2.14 <u>Let</u> $z = \alpha_W(x,y)$, <u>where</u> α_W <u>denotes the group law defining</u> $\hat{W}^+$ <u>on the model</u> $D^{(\underline{\underline{P}})}$, <u>over</u> $\underline{\underline{Z}}$ (1.15). <u>Then</u>, <u>for any</u> $n \in \underline{\underline{P}}$ z_n <u>is given by a polynomial in the</u> x_d , y_d , <u>where</u> $d|n$, <u>with integral coefficients</u>. <u>This poynomial is isobaric of weight</u> n, <u>when</u> x_d, y_d <u>receive the weight</u> d .

2.15 From this, it results that the formal group $\hat{W}^+$ has many factor formal groups. More precisely, <u>let</u> $\mathfrak{N}$ <u>be a subset of</u> $\underline{\underline{P}}$ <u>such that</u> $n \in \mathfrak{N}$ <u>and</u> $d|n$ <u>imply</u> $d \in \mathfrak{N}$. Then we obtain <u>a group law on the model</u> $D^{(\mathfrak{N})}$, just by keeping the components <u>of</u> α_W <u>of index</u> $n \in \mathfrak{N}$, and disregarding all the other ones.

As an important instance, let us take for $\mathfrak{N}$ the set of all integral powers of some prime p, and let us rewrite formulas (2.12) by putting $\xi_h = x_{p^h}$, $\eta_h = y_{p^h}$, $\zeta_h = z_{p^h}$, with $h \in \underline{\underline{N}}$. We obtain, for any $k \in \underline{\underline{N}}$,

2.16 $$\Sigma_{0 \leq k \leq h}\, p^k(\xi_k^{p^{h-k}} + \eta_k^{p^{h-k}}) = \Sigma_{0 \leq h \leq k}\, p^k \zeta_k^{p^{h-k}}$$

2.17 The reason why there is a signe "+" in W^+ and $\hat{W}^+$ is that there is a ring structure on $K^{\underline{\underline{P}}}$ and on $D^{(\underline{\underline{P}})}$: see, later, (IV.4.4).

3. The F_n operators

3.1 Let G be a formal group over a basic ring K, $\gamma \in \mathfrak{C}(G)$ a curve, and $n \in \underline{\underline{P}}$. Then we have a morphism $\sigma_{\gamma,n}: D^n \to G$, defined by

3.2 $$\sigma_{\gamma,n}(t_1,\ldots,t_n) = \gamma(t_1) +\ldots+ \gamma(t_n)$$

or, equivalently

3.3 $$\sigma_{\gamma,n} = \Sigma_{1\leq i\leq n} \gamma\circ p_i ,$$

where $p_i : D^n \to D$ is the i-th coordinate morphism on D^n ; the sums on the right sides are computed in the group $\mathfrak{M}(D^n,G)$.

Now, the morphism $\sigma_{\gamma,n}: D^n \to G$ is obviously symmetric, so that we can apply the symmetric morphism theorem (I.11.19). We obtain a unique factorization

3.4 $$\sigma_{\gamma,n} = s_{\gamma,n} \circ \mathrm{sym}_n ,$$

where $\mathrm{sym}_n: D^n \to D^n$ is as in (I.11.18) and $s_{\gamma,n}: D^n \to G$ is well defined.

We define the morphism $\iota_n: D \to D^n$ by

3.5 $$\iota_n(t) = (0,0,\ldots,(-1)^{n-1}t) ,$$

i.e. $p_i \circ \iota_n = 0$ for $1\leq i<n$ and $(p_n \circ \iota_n)(t) = (-1)^{n-1}t$.

Finally, we define the curve $F_n\cdot\gamma : D \to G$ by putting

3.6 $$F_n\cdot\gamma = s_{\gamma,n} \circ \iota_n .$$

The situation is summarized by the following commutative diagram:

3.7 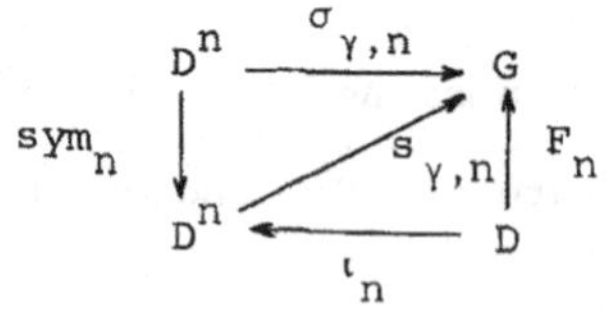

Before discussing the properties of F_n, we will show how to compute $F_n \cdot \gamma$. We have always $F_1 \cdot \gamma = \gamma$.

Let $a_1, \ldots, a_n$ be n elements from K. Then

3.8 $$\Sigma_{1 \leq i \leq n} [a_i] \cdot \gamma = \sigma_{\gamma,n} \circ f ,$$

where $f : D \to D^n$ is the morphism

3.9 $$f(t) = (a_1 t, \ldots, a_n t) .$$

This follows immediately from (3.2) and the definition of [c]: see (I.10.13).

Now, by definition of sym_n (see I.11.18), we have

3.10 $$(\text{sym}_n \circ f)(t) = (b_1 t,\ b_2 t^2, \ldots, b_n t^n) ,$$

where the b_i are related to the a_i by the following equality in $K[X]$:

3.11 $$\Pi_{1 \leq i \leq n} (1 + a_i X) = 1 + \Sigma_{1 \leq i \leq n} b_i X^i$$

Let us assume that K contains n-th roots of unity; more precisely that there are elements $\zeta_1, \ldots, \zeta_n \in K$ such that

3.12 $$\Pi_{1 \leq i \leq n} (X - \zeta_i) = X^n - 1 \quad \text{in} \quad K[X] ,$$

or equivalently

$$\Pi_{1 \leq i \leq n} (1 + \zeta_i X) = 1 + (-1)^{n-1} X^n .$$

Then we obtain from (3.4), (3.8), (3.10), the relation

3.13 $$\Sigma_{1 \leq i \leq n} [\zeta_i] \cdot \gamma = s_{\gamma,n} \circ j_n ,$$

where the ζ_i are as in (3.12) and j_n denotes the morphism

3.14 $$j_n(t) = (0, 0, \ldots, (-1)^{n-1} t^n) .$$

Comparing with (3.6), we obtain

3.15 $$V_n F_n \cdot \gamma = \Sigma_{1 \leq i \leq n} [\zeta_i] \cdot \gamma .$$

3.16 <u>Computation of</u> F_n <u>in additive groups</u>. Let L^+ be an additive formal group (see II.2.21), where any curve γ is written uniquely as

3.17 $$\gamma(t) = \Sigma_{m\in\underline{\underline{P}}} a_m t^m , \quad a_m \in L .$$

Then we have

$$(\Sigma_{1\leq i\leq n} [\zeta_i]\cdot\gamma)(t) = \Sigma_{m\in\underline{\underline{P}}} (\Sigma_{1\leq i\leq n} \zeta_i^m) a_m t^m .$$

But
$$\Sigma_{1\leq i\leq n} \zeta_i^m = \begin{cases} n & \text{if } n|m \\ 0 & \text{if } n\nmid m \end{cases} ,$$

so that

$$(V_n F_n\cdot\gamma)(t) = \Sigma_{m\in\underline{\underline{P}}} na_{nm} t^{nm} ,$$

and

3.18 $$(F_n\cdot\gamma)(t) = \Sigma_{m\in\underline{\underline{P}}} na_{nm} t^m .$$

For the <u>additive group</u> $\underline{\underline{G}}_a$ and its curve $\gamma_{\underline{\underline{a}}}$ (see 1.3), we obtain

3.19 $$F_n\cdot\gamma_{\underline{\underline{a}}} = 0 \quad \underline{\text{for any}} \quad n > 1 .$$

In the <u>multiplicative group</u> $\underline{\underline{G}}_m$, the curve $\gamma = \Sigma_{1\leq i\leq n}[\zeta_i]\cdot\gamma_m$ is given by
$$1-\gamma(t) = \Pi_{1\leq i\leq n} (1-\zeta_i t) = 1 - t^n ,$$
so that

3.20 $$F_n\cdot\gamma_{\underline{\underline{m}}} = \gamma_{\underline{\underline{m}}} , \quad \underline{\text{for any}} \quad n \in \underline{\underline{P}}$$

The formal Witt group $\hat{W}^+$ is defined by the group law α_W on the model $D^{(\underline{\underline{P}})}$ (see 1.15), where we have the basic set of canonical curves ε_i, $i \in \underline{\underline{P}}$ (see II.7.1).

3.21 <u>Definition</u>. <u>We denote by</u> γ_W <u>the curve</u> ε_1 <u>in</u> $\mathbb{C}(\hat{W}^+)$, <u>and we call</u> γ_W <u>the canonical curve of</u> $\hat{W}^+$. Any curve $\gamma \in \mathbb{C}(\hat{W}^+)$ is given by its components $\gamma_i \in \mathbb{C}(D)$, and a relation $\gamma + \gamma' = \gamma''$ in $\mathbb{C}(\hat{W}^+)$ is equivalent to the following equality of formal series (see 1.10):

$$\Pi_{i\in P}(1-\gamma_i(t)\theta^i)(1-\gamma_i'(t)\theta^i) = \Pi_{i\in \underline{\underline{P}}}(1-\gamma_i''(t)\theta^i) .$$

From this, we see that the curve $\Sigma_{1\leq i\leq n}[\zeta_i]\cdot\gamma_W$ corresponds to the "formal series" $(1-t^n\theta^n)$, _so that_ $F_n\cdot\gamma_W$ _is the basic curve_ ε_n. We have proved the following result.

3.22 _Theorem_. In $\mathbb{C}(\hat{W}^+)$, _the set of curves_ $(F_n\cdot\gamma_W)_{n\in\underline{\underline{P}}}$ _is a basic set_. _Moreover, for any_ $A \in \underline{\underline{nil}}(K)$ _and any_ $x = (x_n)_{n\in\underline{\underline{P}}} \in A^{(\underline{\underline{P}})}$, _we have_

3.23 $$x = \Sigma_{n\in\underline{\underline{P}}} (F_n\cdot\gamma_W)(x_n) \quad \text{in the group } \hat{W}^+(A) .$$

Formulas (3.18), (3.19), (3.20), as well as theorem (3.22) have been proved under the assumption that the basic ring K contains n-th roots of unity. The following theorem will allow us to get rid of this unnecessary assumption.

3.24 _Theorem_. _For any formal group_ G _and any_ $n \in \underline{\underline{P}}$, _the map_ $F_n: \mathbb{C}(G) \to \mathbb{C}(G)$ _is additive and continuous_. _Moreover, each_ F_n _commutes to formal group homomorphisms and to changes of rings_.

Proof. Suppose that $\gamma + \gamma' = \gamma''$ in $\mathbb{C}(G)$. Then $\sigma_{\gamma,n} + \sigma_{\gamma',n} = \sigma_{\gamma'',n}$ in the group $\mathfrak{M}(D^n,G)$: see (3.2). From the unicity of $s_{\gamma,n}$ in (3.4), we deduce that $s_{\gamma,n}+s_{\gamma',n} = s_{\gamma'',n}$ in $\mathfrak{M}(D^n,G)$. Then the definition (3.6) gives $F_n\cdot\gamma + F_n\cdot\gamma' = F_n\cdot\gamma''$ in $\mathbb{C}(V)$, i.e. we have proved the additivity of F_n .

Let $u : G \to G'$ be a formal group homomorphism, $\gamma \in \mathbb{C}(G)$ and $\gamma' = u\circ\gamma = \mathbb{C}(u)\cdot\gamma$. Then we obtain successively $\sigma_{n,\gamma'} = u\circ\sigma_{n,\gamma}$, $s_{n,\gamma'} = u\circ s_{n,\gamma}$ and $F_{n\cdot\gamma'} = u\circ(F_n\cdot\gamma)$, i.e.

3.25 $$F_n\cdot(\mathbb{C}(u)\cdot\gamma) = \mathbb{C}(u)\cdot(F_n\cdot\gamma)$$

Let $\varphi : K \to K'$ be a ring homomorphism, G a formal group over K and $\gamma \in \mathbb{C}(G)$. We apply the functor φ_* (see II.2.11), and we write $\gamma^* = \varphi_*\gamma \in \mathfrak{C}(\varphi_*G)$. We have $\varphi_*\sigma_{n,\gamma} = \sigma_{n,\gamma^*}$, $\varphi_* s_{n,\gamma} = s_{n,\gamma^*}$, and $F_n \cdot \gamma^* = \varphi_*(F_n \cdot \gamma)$, i.e.

3.26 $$F_n \cdot (\varphi_*\gamma) = \varphi_*(F_n \cdot \gamma) .$$

Now, if K does not contain n-th roots of unity, it is always possible to find an embedding $\varphi : K \to K'$, where K' contains n-th roots of unity. As φ_* is injective (see I.11.16), we see that formula (3.15) can be applied to $\varphi_*\gamma \in \mathfrak{C}(\varphi_*G)$. This shows the general validity of (3.18), (3.19), (3.20), (3.22). Furthermore, (3.15) gives

$$\operatorname{ord}(V_n F_n \cdot \varphi_*\gamma) \geqslant \operatorname{ord}(\varphi_*\gamma) ,$$

whence $$\operatorname{ord}(F_n \cdot \varphi_*\gamma) \geqslant \frac{1}{n} \operatorname{ord}(\varphi_*\gamma) ,$$

because $\operatorname{ord}(V_n \cdot \gamma) = n \operatorname{ord}(\gamma)$ for any curve γ, and, from (3.26) with injective φ,

3.27 $$\operatorname{ord}(F_n \cdot \gamma) \geqslant \frac{1}{n} \operatorname{ord}(\gamma) ,$$

because $\operatorname{ord}(\varphi_*\gamma) = \operatorname{ord}(\gamma)$: see (I.11.1). Formula (3.27) ends the proof of theorem (3.24), by showing the F_n is continuous.

We see from (3.18) that formula (3.27) is the "best possible" and we stress that <u>the F_n operators are not defined for buds</u>.

4. <u>The representation theorem: $\mathfrak{C}(G) \simeq \operatorname{Hom}(\hat{W}^+, G)$</u>

4.1 <u>Theorem</u>. <u>Let</u> G <u>be a formal group over some basic ring</u>, <u>and</u> $\gamma \in \mathfrak{C}(G)$ <u>a curve</u>. <u>Then there is exactly one formal group homomorphism</u>, $u_\gamma : \hat{W}^+ \to G$, <u>such that</u>

4.2 $$\gamma = u_\gamma \circ \gamma_W \qquad \text{(see 3.21)} .$$

Proof. Let us rewrite formula (3.23) as follows:

4.3 $$\mathrm{Id}_{\hat{W}^+} = \Sigma_{n \in \underline{\underline{P}}} F_n \cdot \gamma_W \circ p_n .$$

The infinite series converges in $\mathfrak{M}(\hat{W}^+, \hat{W}^+)$ for the simple, not for the order topology; p_n denotes the n-th coordinate morphism $D^{(\underline{\underline{P}})} \to D$.

Now we apply to both sides of (4.3) a formal group homomorphism, $u_\gamma \in \mathrm{Hom}(\hat{W}^+,G)$. According to (3.25), we obtain

4.4 $$u_\gamma = \Sigma_{n\in \underline{\underline{P}}} F_n \cdot \gamma \circ p_n ,$$

where $\gamma = u_\gamma \circ \gamma_W$. This shows that u_γ is determined by γ. Furthermore, formula (4.4) defines a <u>morphism</u> $u_\gamma \in \mathfrak{M}(\hat{W}^+,G)$, for any curve γ in any formal group G. It remains to prove that this morphism u_γ is indeed a <u>homomorphism</u>.

Let us first assume that G is an additive group, say $G = L^+$. Then a curve $\gamma \in \mathfrak{C}(G)$ is written uniquely as $\Sigma_{m\in \underline{\underline{P}}} a_m t^m$, $a_n \in L$ (some free K-module), and $F_n\gamma$ is written as $\Sigma_{m\in \underline{\underline{P}}} na_{nm} t^m$. Collecting the "scalars" which multiply a given $a_n \in L$, we translate (4.4) as

4.5 $$u_\gamma = \Sigma_{n\in P} a_n w_n ,$$

where $w_n \in \mathrm{Hom}(\hat{W}^+,K^+)$, as in (2.9). Formula (4.5) shows that u_γ is indeed a homomorphism.

Now let $\varphi : K \to K'$ be a ring homomorphism. It results from (3.26) and the definition (4.4) of the morphism u_γ that

4.6 $$\varphi_* u_\gamma = u_{\varphi_* \gamma} ,$$

for any curve γ in any formal group G over K.

The condition for u_γ to be a homomorphism can be written as

4.7 $$u_\gamma \circ \alpha_W = \mu_G \circ (u_\gamma \times u_\gamma) \qquad \text{(see II.1.9)} .$$

4.8 Applying φ_* to both sides of (4.7), we have, according to (I.11.16), the following proposition. If $\varphi : K \to K'$ is injective and theorem (4.1) holds for $\varphi_* G$, then it holds for G; if φ is surjective and theorem (4.1) holds for G, then it holds for $\varphi_* G$.

4.9 According to the $\underline{\underline{Q}}$ theorem (II.3.1), (4.1) holds when K is a $\underline{\underline{Q}}$-algebra. Then we prove it when K is torsion free, by embedding K in $\underline{\underline{Q}} \otimes K$. Finally we prove it in the general case, according to the lift theorem (II.3.8), and we have established the power of ghosts (see II.3.9).

4.10 Remark. It follows from (4.4) that $\mathrm{ord}(u_\gamma) = \min_{n\in \underline{\underline{P}}} \mathrm{ord}(F_n \cdot \gamma)$. So one can have $\mathrm{ord}(u_\gamma) < \mathrm{ord}(\gamma)$ for some curves γ.

5. Introducing the ring of operators Cart(K)

5.1 In this section, we keep the basic ring K fixed: the formal groups and their homomorphisms are defined over K.

5.2 Definition. An operator (over K) is a natural transformation $x : \mathfrak{C} \to \mathfrak{C}$ on the category of formal groups. In other words, for any formal group G, we have a map

$$x_G : \mathfrak{C}(G) \to \mathfrak{C}(G) ,$$

with the following condition: if $u : G \to G'$ is a formal group homomorphism, we have a commutative diagram, expressed by

$$x_{G'} \circ \mathfrak{C}(u) = \mathfrak{C}(u) \circ x_G .$$

5.3 The composition operators $\mathrm{comp}(\varphi)$, where $\varphi \in \mathfrak{C}(D_K)$ are of this type (they commute to all morphisms). So are the operators F_n (see 3.25). We shall write simply x, not x_G.

5.4 Theorem. Let x, y be two operators. Then $x = y$ iff $x \cdot \gamma_W = y \cdot \gamma_W$. The operators are in one to one correspondence with the curves in $\hat{W}_K^+$. They are all additive.

This follows immediately from theorem (4.1). Indeed, if

5.5 $$\gamma = u_\gamma \circ \gamma_W \quad , \quad \gamma \in \mathfrak{C}(G) \ , \ u_\gamma \in \mathrm{Hom}(\hat{W}^+, G) ,$$

then we must have

5.6 $$x \cdot \gamma = u_\gamma \circ (x \cdot \gamma_W) ,$$

so that x is known once $x \cdot \gamma_W$ is given. Conversely, if $\gamma' \in \mathfrak{C}(\hat{W}^+)$ is any curve, we have an operator defined by the formula

5.7
$$x \cdot \gamma = u_\gamma \circ \gamma' \ .$$

The only property to check is that, for any formal group
homomorphism, $v : G \to G'$ and any $\gamma \in \mathfrak{C}(G)$,

$$u_{v \circ v} = v \circ u_\gamma \ ,$$

which is obvious by composing both sides with γ_w. It follows
from (4.2) that the correspondence $\gamma \mapsto u_\gamma$ is additive, so that
every operator is additive. We shall show later that every operator is continuous (see 7.1).

5.8   The sum $x + y$ and the product xy of two operators are
defined by

$$(x+y) \cdot \gamma = x \cdot \gamma + y \cdot \gamma \quad \text{and} \quad (xy) \cdot \gamma = x \cdot (y \cdot \gamma) \ , \text{ for any } \gamma$$

in any $\mathfrak{C}(G)$. With these definitions, the set of operators becomes a ring.

5.9   Definition. The ring of operators (over K) will be denoted
by Cart(K) , in homage to Pierre Cartier.

5.10   For any formal group G , the additive group $\mathfrak{C}(G)$ has a
structure of left Cart(K)-module. The Cart(K)-module $\mathfrak{C}(\hat{W}^+)$
is free, with one generator, namely γ_w .

This is only a reformulation of previous results.

5.11   From theorem (4.1) applied to $G = \hat{W}^+$, it follows that
$\mathfrak{C}(\hat{W}^+)$ is also a free module, with generator γ_w , over the ring
End$(\hat{W}^+)$ of endomorphisms of $\hat{W}^+$.

5.12   We agree to let End$(\hat{W}^+)$ act on the right, i.e., for any
$u,v \in$ End$(\hat{W}^+)$, uv means u followed by v (not $u \circ v$!). With
this definition, Cart(K) and End$(\hat{W}^+)$ become isomorphic:

$x \in \mathrm{Cart}(K)$ is associated to $u \in \mathrm{End}(\hat{W}^+)$ iff

5.13 $$x \cdot \gamma_W = u \circ \gamma_W .$$

This is easily checked; let us just recall that, if $\mathrm{End}(\hat{W}^+)$ acts on the right on $\hat{W}^+$, then it acts on the left on $\mathrm{Hom}(\hat{W}^+,G) \simeq \mathfrak{C}(G)$.

5.14 Definition. The order of $x \in \mathrm{Cart}(K)$, denoted by $\mathrm{ord}(x)$, is defined by

5.15 $$\mathrm{ord}(x) = \mathrm{ord}(x \cdot \gamma_W) ,$$

or alternatively, according to (5.6)

5.16 $\mathrm{ord}(x) = \inf \mathrm{ord}(x \cdot \gamma)$, for all curves γ in all formal groups.

We have

5.17 $$\mathrm{ord}(x \pm y) \geqslant \min(\mathrm{ord}(x), \mathrm{ord}(y)) ,$$

5.18 $$\mathrm{ord}(xy) \geqslant \mathrm{ord}(x) .$$

5.19 These relations mean that, for any $n \in \underline{\underline{P}}$, the condition $\mathrm{ord}(x) \geqslant n$ defines a right ideal in $\mathrm{Cart}(K)$.

For instance:

5.20 $\mathrm{ord}(V_n) = n$, for any $n \in \underline{\underline{P}}$

5.21 $\mathrm{ord}(F_n) = 1$, for any $n \in \underline{\underline{P}}$

5.22 $\mathrm{ord}([c]) = 1$, for any $c \in K$, $c \neq 0$.

5.23 $\mathrm{ord}(\mathrm{comp}(\varphi)) = \mathrm{ord}(\varphi)$, for any $\varphi \in \mathfrak{C}(D_K)$.

6. Curves in formal groups and representations of operators

6.1 Proposition. Let G be a formal group over K, with a basic set of curves $(\gamma_i)_{i \in I}$. Then any curve $\gamma \in \mathfrak{C}(G)$ can be written uniquely as

6.2 $$\gamma = \Sigma_{m\in\underline{\underline{P}}, i\in I} V_m[x_{m,i}]\cdot\gamma_i ,$$

<u>where the</u> $x_{m,i} \in K$ <u>are subject to the following finiteness condition:</u>

6.3 <u>for any</u> $m \in \underline{\underline{P}}$, $(x_{m,i})_{i\in I} \in K^{(I)}$.

<u>Moreover, if</u>

$$\gamma' = \Sigma_{m\in\underline{\underline{P}}, i\in I} V_m[y_{m,i}]\cdot\gamma_i ,$$

<u>is another curve in</u> G <u>then the relation</u>

6.4 $$\gamma \equiv \gamma' \quad \text{mod.deg. } q$$

<u>is equivalent</u> (<u>for any</u> $q \in \underline{\underline{P}}$) <u>to</u>

6.5 $$x_{m,i} = y_{m,i} \quad \underline{\text{for any}}\ \ m < q \ \ \underline{\text{and any}}\ \ i \in I .$$

<u>If</u> (6.4) <u>holds, then</u>

6.6 $$\text{dif}_q(\gamma,\gamma') = \Sigma_{i\in I}\, \text{dif}_q(V_q\cdot[x_{q,i} - y_{q,i}]\cdot\gamma_i, 0) .$$

In order to prove this proposition, let us denote by C the group $\mathbb{C}(G)$ and, for any $n \in \underline{\underline{P}}$, by C_n the subgroup of C , the elements of which are the γ with $J_{n-1}\,\gamma = 0$.

Note that the relations

6.7 $$\gamma' \equiv \gamma \quad \text{mod } C_n , \text{ i.e. } \gamma' - \gamma \in C_n ,$$

do not depend on the group operation of G, but only on its formal variety structure, because (6.7) is equivalent to

6.8 $$J_{n-1}\,\gamma' = J_{n-1}\,\gamma \quad , \text{ or equivalently } \gamma' \equiv \gamma \quad \text{mod.deg. } n$$

We have the inclusions

$$C = C_1 \supset C_2 \supset \cdots \supset C_n \supset C_{n+1} \supset \cdots .$$

Moreover $\cap_n C_n = 0$ and C is complete for the topology defined by the C_n , i.e. C may be identified with the inverse limit of the discrete groups C/C_n :

6.9 $$C \simeq \varprojlim C/C_n$$

We summarize these properties by saying that C is a <u>filtered complete group</u> (with a filtration indexed by $\underline{\underline{P}}$).

From a more general relation (see I.7.13), we have

6.10 $$V_m \cdot C_n \subset C_{mn} \quad \text{for any } m,n \in \underline{\underline{P}} .$$

If we put

6.11 $$gr_n C = C_n/C_{n+1} \quad \text{for any } n \in \underline{\underline{P}} ,$$

we see from (6.10) that V_m induces maps $gr_n C \to gr_{mn} C$. <u>All these maps are bijective</u>: this becomes obvious when G is an additive group, and we can assume that G is additive, for it does not change anything to the congruences (6.7). In view of the relations $V_{mn} = V_m V_n$, it is sufficient to state that

6.12 <u>the additive map</u> $gr_1 C \to gr_n C$ <u>induced by</u> V_n is bijective for any $n \in \underline{\underline{P}}$.

6.13 <u>The group</u> $gr_1 C = \mathfrak{T}G$ (see I.6.3) <u>is a free</u> K-<u>module</u>. The condition on the basic set $(\gamma_i)_{i\in I}$ is that, when $x = (x_i)_{i\in I}$ ranges over $K^{(I)}$, the curves $\Sigma_{i\in I} [x_i]\cdot\gamma_i$ give a full set of representatives of C_1 modulo C_2.

After these explanations, it should be clear how proposition (6.1) is proved by successive approximations.

6.14 <u>Proposition</u>. <u>Every operator</u> $x \in \mathrm{Cart}(K)$ <u>admits a unique representation in the form</u>

6.15 $$x = \Sigma_{m,n\in\underline{\underline{P}}} V_m[x_{m,n}]F_n , \qquad x_{m,n} \in K ,$$

<u>with the finiteness condition that</u>

6.16 for any $m \in \underline{\underline{P}}$, $(x_{m,n})_{n\in\underline{\underline{P}}} \in K^{(\underline{\underline{P}})}$.

<u>Moreover, if</u> $x \neq 0$, <u>then</u> $\mathrm{ord}(x)$ <u>is the smallest</u> m

for which there is a n *with* $x_{m,n} \neq 0$.

From theorem (5.4) we know that it suffices to show that any curve $\gamma \in \mathbb{C}(\hat{W}^+)$ admits a unique representation of the form $x \cdot \gamma_W$, where x stands for an abbreviation of the right side of (6.15). But in $\mathbb{C}(\hat{W}^+)$ we have the basic set of curves $(F_n \cdot \gamma_W)_{n \in \underline{\underline{P}}}$, according to theorem (3.22). So that proposition (6.1) specializes to proposition (6.14), and more.

7. Continuity, uniform modules, reduced modules

7.1 *Proposition. The ring* Cart(K), *with its order topology (where the right ideals defined by* $\mathrm{ord}(x) \geq n$, $n \in \underline{\underline{P}}$, *are a fundamental system of neighborhoods of* 0) *is a complete filtered ring. The module map* $\mathrm{Cart}(K) \times \mathbb{C}(G) \to \mathbb{C}(G)$ *is continuous for any formal group* G .

Proof. It follows immediately from proposition (6.14) that Cart(K) is filtered and complete qua additive topological group. We have to show that the multiplication map, $\mathrm{Cart}(K) \times \mathrm{Cart}(K) \to \mathrm{Cart}(K)$, is continuous, but this is only a special case of the continuity of a module map, for $\mathbb{C}(\hat{W}^+)$ is a free Cart(K)-module on one generator.

Let $x_o \in \mathrm{Cart}(K)$, $\gamma_o \in \mathbb{C}(G)$ and $n \in \underline{\underline{P}}$ be given. We want to show the existence of $m, m' \in \underline{\underline{P}}$, such that

7.2 $$x \in \mathrm{Cart}(K),\ \mathrm{ord}(x-x_o) \geq m\ ,\ \gamma \in \mathbb{C}(G)\ ,\ \mathrm{ord}(\gamma-\gamma_o) \geq m'$$

imply

7.3 $$\mathrm{ord}(x \cdot \gamma - x_o \cdot \gamma_o) \geq n\ .$$

For this, we write

$$x \cdot \gamma - x_o \cdot \gamma_o = (x-x_o) \cdot \gamma + x_o \cdot (\gamma-\gamma_o)\ ,$$

and we shall insure that both terms on the right side have order

$\geq n$. From (5.16), we know that it suffices to take $m = n$ in order to have $\operatorname{ord}((x-x_o)\cdot\gamma) \geq n$. As concerns the second term, we decompose $x_o \in \operatorname{Cart}(K)$ in the form

$$x_o = x_1 + x_2 \quad , \quad x_1, x_2 \in \operatorname{Cart}(K) \ ,$$

where $\operatorname{ord}(x_1) \geq n$ and x_2 is a <u>finite</u> sum

$$x_2 = \Sigma_{(i,j)\in I} V_i[a_{i,j}]F_j \ , \quad a_{i,j} \in K \ , \ a_{i,j} \neq 0 \ .$$

Proposition (6.14) shows that it is possible. Then we have, for any $\gamma \in \mathfrak{C}(G)$,

$$\operatorname{ord}(F_j\cdot\gamma) \geq \frac{1}{j}\operatorname{ord}(\gamma) \qquad \text{(see 3.27)} \ ,$$

$$\operatorname{ord}([a_{i,j}]F_j\cdot\gamma) \geq \operatorname{ord}(F_j\cdot\gamma) \qquad \text{(see I.7.13)} \ ,$$

$$\operatorname{ord}(V_i[a_{i,j}]F_j\cdot\gamma) \geq i \operatorname{ord}([a_{i,j}]F_j\cdot\gamma) \qquad \text{(see I.7.13)} \ .$$

We conclude that we may take $m' \in \underline{\underline{P}}$ such that

$$m'(\min_{(i,j)\in I} i/j \geq n \ .$$

7.4 <u>Definition</u>. <u>A uniform</u> Cart(K)-<u>module is a topological left</u> Cart(K)-<u>module</u> C <u>having the following property. For any indexed set</u> $(x_j)_{j\in J}$ <u>of elements converging towards</u> 0 <u>in</u> Cart(K), <u>and any set</u> $(\gamma_j)_{j\in J}$ <u>in</u> C , <u>the sum</u>

7.5 $$\Sigma_{j\in J}\, x_j\cdot\gamma_j$$

<u>converges in</u> C .

7.6 Fundamentally, we are interested in characterizing the Cart(K)-modules $\mathfrak{C}(G)$, which have stronger properties than uniformity. Nevertheless, uniform modules will play an auxiliary role (see later, section 10).

7.7 <u>Proposition</u>. <u>Let</u> C <u>be a uniform</u> Cart(K)-<u>module and, for any</u> $n \in \underline{\underline{P}}$, <u>define the additive subgroup</u> C_n <u>of</u> C <u>as the closure of the sum of all subgroups</u> $V_i\cdot C$, <u>for</u> $i \geq n$. <u>Then</u> C <u>is filtered complete for the topology defined by the</u> C_n ,

<u>which is finer than the given topology on</u> C (i.e. any neighborhood of 0 in C contains some C_n) .

Proof. Any element $\gamma_n \in C_n$ can be written in the form

7.8 $$\gamma_n = \Sigma_{i \in I_n} x_i \cdot \gamma_i ,$$

where

7.9 $\text{ord}(x_i) \geq n$ <u>for any</u> $i \in I_n$, <u>and</u> x_i <u>converges towards</u> 0 <u>in</u> C .

Let H be a neighborhood of 0 in C . We have to prove that $\gamma_n \in H$ when n is large enough. We may assume that H is closed. Now, let us assume that, for any $n \in \underline{\underline{P}}$, there is a $\gamma_n \in C_n$, $\gamma_n \notin H$. We can take γ_n as in (7.8), with <u>disjoint</u> indexing sets I_n. Then it follows from (7.9) that the set (x_i), where $i \in I = \cup_n I_n$, converges towards 0 in Cart(K). Therefore, by the uniformity condition, the sum

7.10 $$\Sigma_{i \in I} x_i \cdot \gamma_i$$

converges in C . But this implies that there is a <u>finite</u> subset $I' \subset I$ such that, for any <u>finite</u> subset $I'' \subset I$ not meeting I' (i.e. $I'' \cap I' = \emptyset$),

7.11 $$\Sigma_{i \in I''} x_i \cdot \gamma_i \in H .$$

Now we take $n \in \underline{\underline{P}}$ such that $i \in I_n$ implies $i \notin I'$. From (7.8) and (7.11) we obtain $\gamma_n \in \bar{H} = H$, a contradiction.

As C is a Hausdorff space, we have

7.12 $$\cap_{n \in \underline{\underline{P}}} C_n = 0 .$$

Finally, to show that C is complete for the (C_n) topology, we have to prove that any series $\Sigma_{n \in \underline{\underline{P}}} \gamma_n$, where $\gamma_n \in C_n$ for any $n \in \underline{\underline{P}}$, converges for the (C_n)-topology. But, with our notations (7.8), the sum (7.10) converges towards some

$\gamma \in C$, and we have $\gamma = \Sigma_{n\in \underline{\underline{P}}} \gamma_n$ for the given topology of C (by the general associativity of converging sums). Moreover, $\gamma - \Sigma_{1\leq n<n_o} \gamma_n$ lies in C_{n_o} for any $n_o \in \underline{\underline{P}}$, and this proves that $\gamma = \Sigma_{n\in \underline{\underline{P}}} \gamma_n$ for the (C_n)-topology.

For any uniform Cart(K)-module C we put

7.13 $$gr_n C = C_n/C_{n+1} \quad , \quad n \in \underline{\underline{P}} .$$

<u>Each operator</u> V_m <u>induces an additive map</u> $gr_1 C \to gr_m C$ (and, more generally, maps $gr_n C \to gr_{mn} C$). Moreover $gr_1 C$ <u>has a</u> K-<u>module structure</u>: the multiplication of $gr_1 C$ by $a \in K$ is the map induced by $[a] \in$ Cart(K). This definition makes sense because

7.14 $$\text{ord}([a+b]-[a]-[b]) > 1 \quad \underline{\text{for any}} \quad a,b \in K .$$

According to (5.15), this last relation (7.14) means only that $(a+b)\mathfrak{T}\gamma_W = a\,\mathfrak{T}\gamma_W + b\,\mathfrak{T}\gamma_W$ in the tangent space $\mathfrak{T}\hat{W}^+$.

7.15 <u>Definition</u>. <u>A reduced</u> Cart(K)-<u>module is a uniform</u> Cart(K)-<u>module</u> C , <u>satisfying the following three conditions</u>:

7.16 <u>the topology of</u> C <u>is its</u> (C_n)-<u>topology</u> (see 7.7);

7.17 <u>the map</u> $gr_1 C \to gr_m C$ <u>induced by</u> V_m <u>is bijective</u>, <u>for any</u> $m \in \underline{\underline{P}}$;

7.18 <u>the</u> K-<u>module</u> $gr_1 C$ <u>is free</u> .

We have seen, in (6.12) and (6.13) that any module of curves $\mathfrak{C}(G)$ in a formal group G is a reduced Cart(K)-module. The converse will be called the <u>existence theorem</u>, and will be proved in section (II).

8. How $\mathbb{C}$ is fully faithful on formal groups

Let us consider the following commutative diagram

8.1 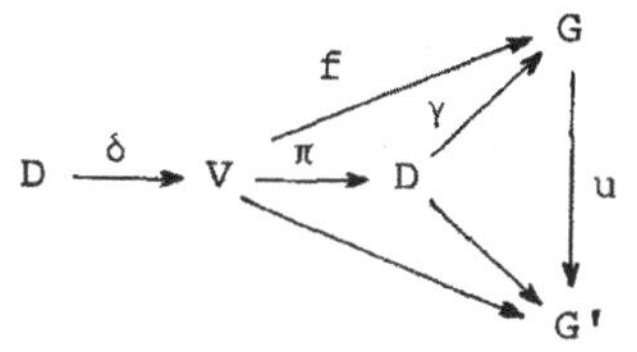

where G and G' denote two given formal groups and u a formal group homomorphism (over some basic ring). As already pointed out (II.1.10), it is indifferent to give u or to give, for any formal variety V , the group homomorphism $\Phi_V : \mathfrak{M}(V,G) \to \mathfrak{M}(V,G')$, with the functorial condition

8.2 $$\Phi_W(f \circ g) = \Phi_V(f) \circ g \ , \text{ for } f \in \mathfrak{M}(V,G) \ , \ g \in \mathfrak{M}(W,V) \ .$$

As $\Phi_V(f) = u \circ f$, we obtain u in the form $u = \Phi_G(Id_G)$.

8.3 Now we assume that u has been lost, but that we are left with Φ_D , i.e. $\mathbb{C}(u)$. From the curves lemma (I.10.3) we know that there is <u>at most one</u> morphism $u : G \to G'$ (a fortiori one homomorphism), with $\mathbb{C}(u) = \Phi_D$.

8.4 <u>Theorem</u>. <u>Let</u> G,G' <u>be two formal groups and</u> $\Phi_D : \mathbb{C}(G) \to \mathbb{C}(G')$ <u>a map</u>. <u>Then there is a formal group homomorphism</u> $u : G \to G'$ <u>such that</u> $\Phi_D(\gamma) = u \circ \gamma$ <u>for any</u> $\gamma \in \mathbb{C}(G)$ <u>iff the map</u> Φ_D <u>is continuous</u>, <u>additive and commutes to the composition operators</u> $comp(\varphi)$, $\varphi \in \mathbb{C}(D)$.

Proof. Continuity of Φ_D and commutation with the $comp(\varphi)$ are implied by the existence of a morphism u, additivity is required for u to be a homomorphism.

Let us assume that Φ_D is given with the above properties. We want to compute $\Phi_V(f)$ for any morphism $f : V \to G$, and we must have

8.5 $$\Phi_V(f)\circ\delta = \Phi_D(f\circ\delta) \quad \text{for any} \quad \delta \in \mathfrak{C}(V) \ .$$

From the curves lemma (I.10.3), we know that there is at most one $\Phi_V(f) \in \mathfrak{M}(V,G')$ satisfying (8.5). If $\Phi_V(f)$ exists, so does $\Phi_W(f\circ g)$ for any morphism $g : W \to V$, and $\Phi_W(f\circ g) = \Phi_V(f)\circ g$.

Now let us choose a basic set $(\gamma_i)_{i\in I}$ in $\mathfrak{C}(G)$. This enables us to write any morphism $f : V \to G$ in the form

8.6 $$f = \Sigma_{i\in I}\ \gamma_i\circ\pi_i$$

with morphisms $\pi_i \in \mathfrak{M}(V,D)$ converging simply towards 0: the π_i are just the components of f for the curvilinear coordinate system associated with the γ_i (see II.7.2). We then define $\Phi_V(f)$ by

8.7 $$\Phi_V(f) = \Sigma_{i\in I}\ \Phi_D(\gamma_i)\circ\pi_i \quad , \quad \text{in} \quad \mathfrak{M}(V,G') \ ,$$

and it only remains to check condition (8.5). Indeed we have

$$\Phi_V(f)\circ\delta = (\Sigma_{i\in I}\ \Phi_D(\gamma_i)\circ\pi_i)\circ\delta$$

$$= \Sigma_{i\in I}\ \Phi_D(\gamma_i)\circ(\pi_i\circ\delta)$$

$$= \Sigma_{i\in I}\ \Phi_D(\gamma_i\circ\pi_i\circ\delta) \quad , \quad \text{because } \Phi_D \text{ commutes}$$

to the composition operators $\mathrm{comp}(\pi_i\circ\delta)$,

$$= \Phi_D(\Sigma_{i\in I}\ \gamma_i\circ\pi_i\circ\delta) \quad , \quad \text{because } \Phi_D \text{ is additive}$$

and continuous,

$$= \Phi_D(f\circ\delta) \ , \quad \text{Q.E.D.}$$

8.8 <u>Remark</u>. We did not assume that Φ_D commutes to the F_n operators. It is a consequence of the existence of $u \in \mathrm{Hom}(G,G')$ with $\Phi_D = \mathfrak{C}(u)$: see (3.25).

9. Some corollaries of the representation theorem

9.1 *Identities with one variable curve*. Let $f = (f_i)_{i\in I}$ be a morphism from a formal variety V into a model $D^{(I)}$, and $(x_i)_{i\in I}$ be any set in $\mathrm{Cart}(K)$ indexed by I. Then the following three relations are equivalent

9.2 $$\Sigma_{i\in I}\ (x_i\cdot\gamma_W)\circ f_i = 0\ ,\quad \text{in}\quad \mathfrak{M}(V,\hat{W}^+)\ ;$$

9.3 $$\Sigma_{i\in I}\ (x_i\cdot\gamma)\circ f_i = 0\ ,\quad \text{in}\quad \mathfrak{M}(V,G),\ \underline{\text{for any curve}}\ \gamma\ \underline{\text{in any formal group}}\ G;$$

9.4 $$\Sigma_{i\in I}\ \mathrm{comp}(f_i\circ\delta)x_i = 0\ ,\quad \text{in}\quad \mathrm{Cart}(K),\ \text{for any curve}\ \delta\in\mathfrak{C}(V).$$

Proof. When applying the group homomorphism $u_\gamma: \hat{W}^+ \to G$, with $u_\gamma\circ\gamma_W = \gamma$ (4.1), to (9.2), we obtain (9.3), which is therefore equivalent to its special instance (9.2). Equivalence of (9.2) and (9.4) comes from the curves lemma (see I.10.3) and theorem (5.4).

9.5 *Let $D^{(J)}$ be a model, with its projections $p_j: D^{(J)} \to D$ and its curves $\varepsilon_j: D \to D^{(J)}$, $j\in J$, as in (II.7.1). Then there are unique morphisms $\sigma_{m,J}: D^{(J)} \to D$, $m\in \underline{\underline{P}}$, such that*

9.6 $$\Sigma_{j\in J}\ \gamma_W\circ p_j = \Sigma_{m\in\underline{\underline{P}}}\ (F_m\cdot\gamma_W)\circ\sigma_{m,J}\ ,\quad \underline{\text{in}}\quad \mathfrak{M}(D^{(J)},\hat{W}^+)\ .$$

Moreover

9.7 $$\sigma_{1,J} = \Sigma_{j\in J}\ p_j\qquad \text{(in the ordinary additive group } \mathfrak{M}(D^{(J)},D)\text{)};$$

9.8 $$\sigma_{m,J}\circ\varepsilon_j = 0\qquad \underline{\text{for any}}\ m\geq 2\ \underline{\text{and any}}\ j\in J\ .$$

Proof. The existence and unicity of the morphisms $\sigma_{m,J}$ comes from (3.22). Formula (9.7) is obtained by substituting $\gamma_{\underline{\underline{a}}}$ to γ_W in (9.6): see (3.19) and (9.1). Composing both sides of (9.6) with ε_j, we obtain, according to (9.7)

$$\gamma_W = \gamma_W + \Sigma_{m\geq 2}\ (F_m\cdot\gamma_W)\circ(\sigma_{m,J}\circ\varepsilon_j)\ ,$$

or equivalently

9.9 $$\Sigma_{m\geq 2} (F_m \cdot \gamma_W) \circ (\sigma_{m,J} \circ \varepsilon_j) = 0 ,$$

which, according to (3.22), is equivalent to (9.8).

9.10 Put, in (9.5), $J = \{1,\dots,n\}$, so that $D^{(J)} = D^n$, and write $\sigma_{m,n}$ instead of $\sigma_{m,J}$. Then $\sigma_{m,n}: D^n \to D$ is a symmetric morphism for any $m \in \underline{\underline{P}}$ and therefore (see I.11.19) there are morphisms $s_{m,n}$, defined uniquely by

9.11 $$\sigma_{m,n} = s_{m,n} \circ \mathrm{sym}_n \quad , \quad \text{for any} \quad m \in \underline{\underline{P}}$$

<u>If the morphism</u> $\iota_n: D \to D^n$ <u>is as in</u> (3.5), <u>then</u>

9.12 $$s_{m,n} \circ \iota_n = \delta_{m,n}\, \mathrm{Id}_D \quad \underline{\text{for any}} \quad m \in \underline{\underline{P}} \text{ , where } \delta_{m,n} = 0$$

or 1 is the Kronecker symbol.

Indeed, when applying the definition of F_n (see 3.6) to the canonical curve $\gamma_W \in \mathbb{C}(\hat{W}^+)$, the morphism $\sigma_{\gamma_W,n}$ is

$$\Sigma_{1\leq i\leq n}\, \gamma_W \circ p_i = \Sigma_{m\in\underline{\underline{P}}} (F_m \cdot \gamma_W) \circ \sigma_{m,n} \; ,$$

and, according to (3.4) and (9.11), $s_{\gamma_W,n}$ is

$$\Sigma_{m\in\underline{\underline{P}}} (F_m \cdot \gamma_W) \circ s_{m,n} \quad ,$$

so that (9.12) is a consequence of definition (3.6) and of (3.22).

9.13 <u>Let</u> $\pi = (\pi_j)_{j\in J} : V \to D^{(J)}$ <u>be a morphism with the following property: for some given</u> $n \in \underline{\underline{P}}$, $\mathrm{ord}(\pi_j) \geq n$ <u>except maybe for one</u> $j \in J$, <u>say</u> j_o . <u>Then</u>

9.14 $$\mathrm{ord}(\sigma_{m,J} \circ \pi) \geq n + \mathrm{ord}(\pi_{j_o}) \geq n + 1 \; , \quad \underline{\text{for any}} \quad m \geq 2.$$

Proof. The condition on π may be reformulated in the following form:

9.15 $$\pi \equiv \varepsilon_{j_o} \circ \pi_{j_o} \qquad \text{mod.deg. } n$$

Then (9.14) can be derived directly from the composition lemma (I.7.1) and relation (9.8), provided we prove

9.16 $$\mathrm{ord}(\sigma_{m,J}) \geq 2 \quad , \quad \text{for any} \quad m \geq 2 \ .$$

But this is equivalent to $\mathfrak{T}\sigma_{m,J} = 0$ (for $m \geq 2$), and comes from (9.8).

<u>We have</u>

9.17 $$\mathrm{ord}(\mathrm{comp}(\varphi+\varphi') - \mathrm{comp}(\varphi) - \mathrm{comp}(\varphi')) \geq \mathrm{ord}(\varphi) + \mathrm{ord}(\varphi') \ ,$$

<u>for any two</u> $\varphi, \varphi' \in \mathfrak{C}(D)$.

Indeed we obtain, from (9.6) and (9.7), when composing with $(\varphi,\varphi') : D \to D^2$, according to theorem (5.4)

9.18 $$\mathrm{comp}(\varphi) + \mathrm{comp}(\varphi') - \mathrm{comp}(\varphi+\varphi') = \Sigma_{m\geq 2} s_{m,2} \circ (\varphi,\varphi') \ ,$$

so that (9.17) comes from (9.13).

10. <u>The existence theorem: preliminaries with a uniform module</u>

10.1 The problem is to reconstruct a formal group G from its Cart(K)-module $\mathfrak{C}(G)$. We know that G is defined by $\mathfrak{C}(G)$, up to isomorphism (see 8.4). But what are the conditions on a Cart(K)-module C in order that there exists a formal group G with a Cart(K)-module isomorphism $C \to \mathfrak{C}(G)$? The answer is: C <u>has to be a reduced</u> Cart(K)-<u>module</u> (see 7.15), but we shall first use only the property that C is a <u>uniform</u> Cart(K)-module (see 7.4).

10.2 More precisely, <u>let</u> C <u>be a uniform</u> Cart(K)-<u>module</u>. We shall show presently how to associate to any formal variety V a group $\Gamma(V)$ and to any morphism $u : W \to V$ a group homomorphism $\Gamma(u) : \Gamma(V) \to \Gamma(W)$. In this way we shall obtain <u>a contravariant functor in groups</u> on the category of formal varieties.

Later we shall show that Γ is representable, under the supplementary hypothesis that C is reduced.

10.3 We must remember that $\Gamma(V)$ is to be identified later with $\mathfrak{M}(V,G)$. We shall use the faithfulness of the functor $\mathfrak{C}$, and replace any $f \in \mathfrak{M}(V,G)$ by $\mathfrak{C}(f): \mathfrak{C}(V) \to \mathfrak{C}(G)$, because $\mathfrak{C}(G)$ is given as C. That is why <u>we define</u> $\Gamma(V)$ <u>as a subgroup of the group of all applications of</u> $\mathfrak{C}(V)$ <u>into</u> C.

Let us consider the diagram

10.4 $$D \xrightarrow{\delta} V \xrightarrow{\pi} D \xrightarrow{\gamma} G .$$

As a matter of fact, it is a wishful diagram, for G is yet unknown. But γ makes sense, as an element of C. So does $\gamma \circ \pi \circ \delta \in C$, written as $\mathrm{comp}(\pi\circ\delta)\cdot\gamma$, for $\mathrm{comp}(\pi\circ\delta) \in \mathrm{Cart}(K)$ and C is given with its $\mathrm{Cart}(K)$-module structure.

10.5 <u>Definition</u>. <u>Let</u> V <u>be a formal variety</u>, $(\gamma_j)_{j\in J}$ <u>any indexed set of elements in</u> C <u>and</u> $(\pi_j)_{j\in J}$ <u>a set of morphisms</u> $V \to D$, <u>converging simply towards</u> 0. <u>In other words</u>, $\pi = (\pi_j)_{j\in J}$ <u>is a morphism</u> $V \to D^{(J)}$ <u>given by its components</u>. <u>Then we define</u>

$$\Sigma_{j\in J} \quad \gamma_j * \pi_j$$

<u>to be the map</u>

10.6 $$f : \delta \mapsto \Sigma_{j\in J} \quad \mathrm{comp}(\pi_j\circ\delta)\cdot \gamma_j \quad ,$$

<u>from</u> $\mathfrak{C}(V)$ <u>into</u> C.

As π_j converges towards 0, so does $\pi_j\circ\delta$ and $\mathrm{comp}(\pi_j\circ\delta)$: see (I.2.8) and (5.23). The existence of f relies only on the uniformity of C.

It is clear from this definition that the maps f defined as above make up <u>a subgroup</u> $\Gamma(V)$ <u>of the group of all maps of</u>

$\mathfrak{C}(V)$ <u>into</u> C .

10.7 For any $n \in \underline{\underline{P}}$, we define the subgroup $\Gamma(V)_n$ of $\Gamma(V)$ as follows : $f \in \Gamma(V)$ <u>lies in</u> $\Gamma(V)_n$ <u>iff</u> f <u>admits a representation</u> (10.6) <u>with</u> $\mathrm{ord}(\pi_j) \geq n$ <u>for any</u> $j \in J$.

10.8 Let $u \in \mathfrak{M}(W,V)$ and $f \in \Gamma(V)$ be given. Then we define $\Gamma(u)(f) = f' \in \Gamma(W)$ by putting

10.9 $$f'(\delta) = f(u \circ \delta) \text{ , for any } \delta \in \mathfrak{C}(W) \text{ .}$$

If f is as in (10.6), this amounts to replacing each $\pi_j : V \to D$ by $\pi_j \circ u : W \to D$. Moreover $\Gamma(u)$ is a filtered group homomorphism (i.e. it sends $\Gamma(V)_n$ into $\Gamma(W)_n$ for any $n \in \underline{\underline{P}}$), and Γ has the required functorial properties: $\Gamma(u \circ v) = \Gamma(v) \circ \Gamma(u)$, $\Gamma(\mathrm{Id}) = \mathrm{Id}$.

10.10 <u>Lemma</u>. <u>Let</u> $\pi : V \to D$ <u>and</u> $\varphi = (\varphi_j)_{j \in J} : D \to D^{(J)}$ <u>be morphisms, and let</u> $(\gamma_j)_{j \in J}$ <u>be a set in</u> C . <u>Then</u>

10.11 $$\Sigma_{j \in J} \gamma_{j*}(\varphi_j \circ \pi) = (\Sigma_{j \in J} \mathrm{comp}(\varphi_j) \cdot \gamma_j) * \pi \text{ .}$$

Proof. We apply the definition (10.6), and we have to prove that

$\Sigma_{j \in J} \mathrm{comp}(\varphi_j \circ \pi \circ \delta) \cdot \gamma_j = \mathrm{comp}(\pi \circ \delta) \Sigma_{j \in J} \mathrm{comp}(\varphi_j) \gamma_j$, which results from (I.10.11) and from the general properties of continuous Cart(K)-modules.

When we specialize lemma(10.10) by putting $V = D$ and $\pi = \mathrm{Id}_D$, we obtain the following result.

10.12 <u>The map</u> $\Sigma_{j \in J} \gamma_j * \varphi_j \mapsto \Sigma_{j \in J} \mathrm{comp}(\varphi_j) \cdot \gamma_j$ <u>is an isomorphism of</u> $\Gamma(D)$ <u>onto</u> C , <u>qua additive groups</u>.

10.13 <u>Lemma</u>. <u>Let</u> $\Sigma_{i \in I} (x_i \cdot \gamma_w) \circ f_i = 0$ <u>be an identity, as in</u> (9.1). <u>Then</u>

10.14 $$\Sigma_{i\in I}\ (x_i\cdot\gamma)_*\ f_i = 0 \quad \text{holds in} \quad \Gamma(V)\ .$$

Proof. We use the equivalence of (9.2) and (9.4) when applying definition (10.6) .

10.15 Lemma. Let $\pi = (\pi_j)_{j\in J} : V \to D^{(J)}$ be a morphism satisfying condition (9.13) , i.e. $\operatorname{ord}(\pi_j) \geq n$, with at most one exception. Then, for any $\gamma \in C$

10.16 $$\Sigma_{j\in J}\ \gamma_*\pi_j - \gamma_*(\Sigma_{j\in J}\ \pi_j) \in \Gamma(V)_{n+1}\ .$$

Proof. We apply lemma (10.13) to the identity

$$\Sigma_{j\in J}\ \gamma_w\circ\pi_j - \gamma_w\circ(\Sigma_{j\in J}\ \pi_j) = \Sigma_{m\geq 2}\ (F_m\cdot\gamma_w)\circ(\sigma_{m,J}\circ\pi)$$

(see 9.5) , then we apply (9.14) and the definition of $\Gamma(V)_{n+1}$ (see 10.7) .

11. The existence theorem: end of the proof

11.1 Now we assume that C is a reduced $\operatorname{Cart}(K)$-module, and we translate definition (7.15) as follows.

11.2 Let us call a V-basis of C an indexed set $(\gamma_i)_{i\in I}$ in C , such that the residual classes modulo C_2 of the γ_i are a basis of the free K-module C/C_2. Then we obtain for any $n \in \underline{\underline{P}}$ a full set of representatives of C_n modulo C_{n+1} by taking the elements

11.3 $$\Sigma_{i\in I}\ V_n[x_i]\cdot\gamma_i\ ,$$

where $(x_i)_{i\in I}$ ranges over $K^{(I)}$.

11.4 Lemma. The choice of a V-basis $(\gamma_i)_{i\in I}$ in C puts C in one-to-one correspondence with the set of curves $\mathfrak{C}(D^{(I)})$. To any curve $\varphi = (\varphi_i)_{i\in I} : D \to D^{(I)}$ there corresponds the element

11.5 $$\Sigma_{i\in I} \operatorname{comp}(\varphi_i)\cdot\gamma_i \in C .$$

<u>This correspondence is compatible with the inverse limits structures of</u> $\mathfrak{C}(D,D^{(I)})$ <u>and of</u> C ; <u>i.e. for any</u> $n \in \underline{\underline{P}}$, <u>the</u> n-<u>jets of curves</u> $\varphi : D \to D^{(I)}$ <u>are in one-to-one correspondence with the elements of</u> C/C_{n+1} .

This lemma is proved by successive approximations, i.e. by induction on $n \in \underline{\underline{P}}$. Let $\gamma \in C$ and $\varphi \in \mathfrak{C}(D^{(I)})$ be given, such that

11.6 $$\gamma - \Sigma_{i\in I} \operatorname{comp}(\varphi_i)\cdot\gamma_i \in C_n .$$

Then there is a unique $(x_i)_{i\in I} \in K^{(I)}$, such that

$$\gamma - \Sigma_{i\in I} (\operatorname{comp}(\varphi_i) + V_n[x_i])\cdot\gamma_i \in C_{n+1} .$$

On the other hand, if $\varphi' \in \mathfrak{C}(D^{(I)})$, $\operatorname{ord}(\varphi') \geq n$, then we have, according to (9.15),

$$\operatorname{ord}(\operatorname{comp}(\varphi_i+\varphi_i') - \operatorname{comp}(\varphi_i) - \operatorname{comp}(\varphi_i')) \geq (n+1) \quad \text{in} \quad \operatorname{Cart}(K),$$

for any $i \in I$; moreover, the relations

$$\varphi_i'(t) \equiv y_i t^n \quad \text{mod.deg. } (n+1),\ i \in I,\ (\varphi_i)_{i\in I} \in K^{(I)} ,$$

imply

$$\operatorname{ord}(\operatorname{comp}(\varphi_i') - V_n[y_i]) \geq n+1 \quad \text{in} \quad \operatorname{Cart}(K) .$$

This shows that, by putting $\varphi_i'(t) = x_i t^n$, we have

11.7 $$\gamma - \Sigma_{i\in I} \operatorname{comp}(\varphi_i+\varphi_i')\cdot\gamma_i \in C_{n+1} ,$$

and that the n-jet of $\varphi+\varphi'$ is determined by (11.7), provided that the (n-1)-jet of φ is determined by (11.6) .

11.8 <u>Proposition</u>. <u>The choice of a</u> V-<u>basis</u> $(\gamma_i)_{i\in I}$ <u>in</u> C <u>determines, for any formal variety</u> V , <u>a homeomorphism</u>

$$\eta_V : \mathfrak{M}(V,D^{(I)}) \to \Gamma(V) ,$$

<u>by the formula</u>

11.9 $$\eta_V(\pi) = \Sigma_{i\in I}\ \gamma_i * \pi_i \quad , \quad \text{for} \quad \pi = (\pi_i)_{i\in I} : V \to D^{(I)} \ ,$$

<u>when</u> $\mathfrak{M}(V,D^{(I)})$ <u>is endowed with its order topology and</u> $\Gamma(V)$ <u>with its filtration topology.</u>

<u>The maps</u> η_V <u>are functorial, i.e.:</u>

11.10 $$\eta_W(\pi \circ u) = \Gamma(u)(\eta_V(\pi))$$

<u>for any morphism</u> $u : W \to V$ <u>of formal varieties, and any</u> $\pi \in \mathfrak{M}(V,D^{(I)})$.

Proof. We take (11.9) as a definition of the map η_V. Then (11.10) follows from the definition of $\Gamma(u)$: see (10.8).

A relation $\eta_V(\pi) = \eta_V(\pi')$, where $\pi, \pi' \in \mathfrak{M}(V,D^{(I)})$, means, by definition (see 10.6),

$$\Sigma_{i\in I}\ \mathrm{comp}(\pi_i \circ \delta)\cdot\gamma_i = \Sigma_{i\in I}\ \mathrm{comp}(\pi'_i \circ \delta)\cdot\gamma_i \quad \text{for any} \quad \delta \in \mathfrak{C}(D),$$

so that it implies $\pi \circ \delta = \pi' \circ \delta$ by lemma (11.4), and $\pi = \pi'$ by the curves lemma. We have proved that η_V is <u>injective</u>.

We shall show that η_V is surjective by successive approximations, using an "expanding and collecting" process. Let $f \in \Gamma(V)$ and $n \in \underline{\underline{P}}$ be given, such that

11.11 $$f - \Sigma_{i\in I}\ \gamma_i * \pi_i \in \Gamma(V)_n \ .$$

This means that there is some morphism $\psi = (\psi_j)_{j\in J'} : V \to D^{(J')}$, with

11.12 $$\mathrm{ord}(\psi) = \min_{j\in J'} \mathrm{ord}(\psi_j) \geq n \ , \quad \text{and}$$

11.13 $$f - \Sigma_{i\in I}\ \gamma_i * \pi_i = \Sigma_{j\in J'}\ \gamma_j * \psi_j \ ,$$

for some set $(\gamma_j)_{j\in J'}$ in C (we write J', not J, because we shall presently define J by adjoining one element to the set J').

According to lemma (11.4), we expand each γ_j in the form

11.14 $$\gamma_j = \Sigma_{i\in I}\ \mathrm{comp}(\varphi_{i,j})\cdot\gamma_i ,$$

and we apply lemma (10.10) to write, for any $j \in J'$

11.15 $$\gamma_j * \psi_j = \Sigma_{i\in I}\ \gamma_i * (\varphi_{i,j} \circ \psi_j) .$$

Note that the morphisms $\varphi_{i,j} \circ \psi_j$ converge simply towards 0 in $\mathfrak{M}(V,D)$: indeed for any $A \in \underline{\underline{\mathrm{nil}}}(K)$ and any $x \in V(A)$, we have $\psi_{j,A}(x) = 0$ for almost all $j \in J'$ and, for any $j \in J'$, $\varphi_{i,j,A}(\psi_{j,A}(x)) = 0$ for almost all $i \in I$. So we may substitute the right side of (11.15) for $\gamma_j * \psi_j$ in (11.13) and intervert summations. We obtain

11.16 $$f = \Sigma_{i\in I}\ (\gamma_i * \pi_i + \Sigma_{j\in J'}\ \gamma_i * (\varphi_{i,j} \circ \psi_j)) .$$

Put $J = J' \circ \{j_o\}$ for some $j_o \notin J'$, and

11.17 $$\pi_i' = \pi_i + \Sigma_{j\in J'}\ (\varphi_{i,j} \circ \psi_j) , \quad \pi' = (\pi_i')_{i\in I} .$$

We have $\pi' \equiv \pi$ mod.deg. n, from (11.12), and we are in position to apply lemma (10.15), separately for any $i \in I$ (we find the very notations of this lemma by putting $\pi_{j_o} = \pi_i$ and $\pi_j = \varphi_{i,j} \circ \psi_j$ for $j \in J'$). We obtain, from (11.16),

11.18 $$f - \eta_V(\pi') \in \Gamma(V)_{n+1} .$$

Note that $\cap_{n\in\underline{\underline{P}}}\ \Gamma(V)_n = 0$, because $f \in \Gamma(V)_n$ implies $f(\delta) \in C_n$ for any $\delta \in \mathfrak{C}(V)$, from definition (10.6) and lemma (11.4). So we have proved the <u>surjectivity</u> of η_V, and that $\eta_V(\pi) - \eta_V(\pi') \in \Gamma(V)_n$ implies $\pi \equiv \pi'$ mod.deg. n. Conversely, if $\pi \equiv \pi'$ mod. deg. n, we put $\pi' - \pi = \pi''$, and lemma (10.15) shows that

$$\gamma_i * \pi_i + \gamma_i * \pi_i'' - \gamma_i * \pi_i' \in \Gamma(V)_{n+1} ,$$

with $\gamma_i * \pi_i'' \in \Gamma(V)_n$, so that $\eta_V(\pi) - \eta_V(\pi') \in \Gamma(V)_n$. This completes the proof of proposition (11.8).

Proposition (11.8) says that the choice of a V-basis (γ_i) in C leads to a representation of the contravariant functor in groups Γ on the model $D^{(I)}$. More precisely, there is a curvilinear group law μ on the model $D^{(I)}$, such that, for any formal variety V,

$$\eta_V : \mathfrak{M}(V,G) \to \Gamma(V)$$

is an <u>isomorphism of topological groups</u>; of course, $\mathfrak{M}(V,G)$ is endowed with the group structure defined by μ. As for $\mu : D^{(I)} \times D^{(I)} \to D^{(I)}$, it can be defined by the relation

11.19 $$\eta_V(\pi) + \eta_V(\pi') = \eta_V(\mu(\pi,\pi')) ,$$

for any formal variety V and any two morphisms $\pi,\pi' : V \to D^{(I)}$, or, alternatively, by

11.20 $$\Sigma_{i\in I} (\mathrm{comp}(\delta_i) + \mathrm{comp}(\delta'_i)\cdot\gamma_i = \Sigma_{i\in I} \mathrm{comp}(\mu(\delta,\delta')_i)\cdot\gamma_i ,$$

for any two curves $\delta,\delta' \in \mathfrak{C}(D^{(I)})$.

There is still a point to be checked. Is $\mathfrak{C}(G)$ isomorphic to C ? From (10.12), (11.4) and (11.8) we have established an isomorphism between $\mathfrak{C}(G)$ and C qua topological groups. This isomorphism commutes to the operators V_n and $[c]$: this is a consequence of the functoriality of η in (11.8), because V_n and $[c]$ are composition operators, but what about F_n ? In other words, if $\gamma_G \in \mathfrak{C}(G)$ corresponds to $\gamma \in C$, does $F_n\cdot\gamma_G \in \mathfrak{C}(G)$ correspond to $F_n\cdot\gamma \in C$? We have to go back to definition (3.6).

From the properties of η, we have

11.21 $$\eta_{D^n}(\Sigma_{1\leq i\leq n} \gamma_G \circ p_i) = \Sigma_{1\leq i\leq n} \gamma * p_i \quad \text{in} \quad \Gamma(D^n) .$$

From lemma (10.15) and the identity

$$\Sigma_{1\leq i\leq n}\ \gamma_W \circ p_i = \Sigma_{m\in \underline{\underline{P}}}\ (F_m \cdot \gamma_W)\circ(s_{m,n}\circ sym_n)\ ,$$

we obtain

$$\Sigma_{1\leq i\leq n}\ \gamma * p_i = \Sigma_{m\in \underline{\underline{P}}}\ (F_m\cdot\gamma) * (s_{m,n}\circ\ sym_n) \quad \text{in} \quad \Gamma(D^n)\ ,$$

so that

11.22 $$\eta_{D^n}(\Sigma_{1\leq i\leq n}\ \gamma_G \circ p_i) = \Sigma_{m\in \underline{\underline{P}}}(F_m\cdot\gamma) * (s_{m,n}\circ\ sym_n) \quad \text{in} \quad \Gamma(D^n)$$

But $\Gamma(sym_n)$ corresponds to composition with sym_n, and is therefore <u>injective</u>, so that (11.22) implies, with the notations of (3.4),

11.23 $$\eta_{D^n}(s_{\gamma_G,n}) = \Sigma_{m\in \underline{\underline{P}}}(F_m\cdot\gamma) * s_{m,n}\ .$$

Composing with ι_n (3.5) and applying (9.12), we obtain

11.24 $$\eta_D(F_n\cdot\gamma_G) = F_n\cdot\gamma \quad , \text{ as required.}$$

So, <u>for any reduced</u> Cart(K)-<u>module</u> C , <u>there is a formal group</u> G <u>such that</u> $\mathfrak{C}(G)$ <u>is isomorphic to</u> C , <u>qua reduced</u> Cart (K)-<u>module</u>. This is the existence theorem, and we can state our results as follows.

11.25 <u>Theorem</u>. <u>The category of the formal groups over a basic ring</u> K <u>is equivalent to the category of the reduced</u> Cart(K)-<u>modules</u>, when one takes formal group homomorphisms as morphisms in the first category, and continuous Cart(K)-linear maps as morphisms in the second one.

CHAPTER IV

THE SPECIAL EQUIVALENCES OF CATEGORIES

1. The functor Cart and the commuting relations

1.1 So far, we have only considered the ring of operators Cart(K) for a fixed basic ring K . Now let $\varphi : K \to K'$ be a ring homomorphism and G a formal group over K . Then we have the formal group $\varphi_* G$ over K' , and the morphism of topological groups $\varphi_* : \mathfrak{C}(G) \to \mathfrak{C}(\varphi_* G)$. (see II.2.11.) For any curve $\gamma \in \mathfrak{C}(G)$, we have

1.2 $$\varphi_*(V_n \cdot \gamma) = V_n \cdot \varphi_* \gamma \quad , \quad \text{for any } n \in \underline{\underline{P}} \quad \text{(see I.11.11)} \; ;$$

1.3 $$\varphi_*([c] \cdot \gamma) = [\varphi(c)] \cdot \varphi_* \gamma \quad , \quad \text{for any } c \in K \quad \text{(see I.11.12)} \; ;$$

1.4 $$\varphi_*(F_n \cdot \gamma) = F_n \cdot \varphi_* \gamma \; , \quad \text{for any } n \in \underline{\underline{P}} \quad \text{(see III.3.26)} \; .$$

The above formulas lay bare the functorial character of Cart(K) , and can be summarized as follows (see III.6.14).

1.5 Corresponding to every basic ring homomorphism $\varphi : K \to K'$, there is a ring homomorphism $\varphi_* : \mathrm{Cart}(K) \to \mathrm{Cart}(K')$, which maps

$$x = \Sigma_{m,n \in \underline{\underline{P}}} V_m [x_{m,n}] F_n \in \mathrm{Cart}(K)$$

on

$$\varphi_* x = \Sigma_{m,n \in \underline{\underline{P}}} V_m [\varphi(x_{m,n})] F_n \in \mathrm{Cart}(K') \; .$$

For any curve in any formal group over K , say $\gamma \in \mathfrak{C}(G)$, we have.

1.6 $$\varphi_*(x \cdot \gamma) = \varphi_* x \cdot \varphi_* \gamma \; .$$

As our purpose is to replace Cart(K) by a more manageable ring $\mathrm{Cart}_S(K)$, where S denotes a set of primes, the functor Cart will become no more than a special case of Cart_S (namely, when $S = P$, the set of all primes). Nevertheless we need to prove the following formulas, for $m,n \in \underline{\underline{P}}$, $c \in K$:

1.7
$$\begin{cases} F_m F_n = F_{mn} & ; \\ F_n[c] = [c^n]F & ; \\ F_n V_n = n\cdot 1_{Cart(K)} & ; \\ \text{if } \gcd(m,n) = 1, \text{ then } V_m F_n = F_n V_m & . \end{cases}$$

1.8 Let us compute the action of Cart(K) on $\mathbb{C}(\underline{\underline{G}}_a)$. We remember that $\mathbb{C}(\underline{\underline{G}}_a)$ is just the ordinary additive group of power series $\Sigma_{i \in \underline{\underline{P}}} a_i t^i$, $a_i \in K$, so that the additive operators $x \in$ Cart(K) act additively on $\mathbb{C}(\underline{\underline{G}}_a)$ in the plain sense (see II.2.19). As $a_i t^i \in \mathbb{C}(\underline{\underline{G}}_a)$ is mapped by V_n on $a_i t^{ni}$, by $[c]$ on $c^i a_i t^i$, by F_n on 0, unless n/i and then on $na_i t^{i/n}$ (see III.3.16), we obtain a matrix representation of Cart(K), which will presently be made explicit.

1.9 Definition. For any indexing set I, the ring of linear continuous endomorphisms of the topological K-module K^I (where K has the discrete topology) is identified with the ring of square matrices with finite rows, denoted by $M_{(I)}(K)$. An element $x \in M_{(I)}(K)$ is a matrix $(x_{i,j})_{i,j \in I}$ with entries $x_{i,j} \in K$ verifying the finiteness condition:

for any $i \in I$, $(x_{i,j})_{j \in I} \in K^{(I)}$.

Such an x maps $a = (a_i)_{i \in I} \in K^I$ on $a' = (a'_i)_{i \in I}$, where $a'_i = \Sigma_{j \in I} x_{i,j} a_j$, and will be also written as $\Sigma_{i,j \in I} x_{i,j} \underline{\underline{e}}_{i,j}$, by using the matrix units $\underline{\underline{e}}_{i,j}$.

1.10 Proposition. The action of Cart(K) on $\mathbb{C}(\underline{\underline{G}}_a)$, identified with $K^{\underline{\underline{P}}}$, is given by the functorial ring homomorphism κ : Cart(K) $\rightarrow M_{(\underline{\underline{P}})}(K)$, defined as follows

1.11
$$\begin{cases}\kappa(V_m) = \Sigma_{i\in\underline{\underline{P}}}\ \underline{\underline{e}}_{mi,i} ;\\ \kappa([c]) = \Sigma_{i\in\underline{\underline{P}}}\ c^i \underline{\underline{e}}_{i,i} ;\\ \kappa(F_n) = \Sigma_{i\in\underline{\underline{P}}}\ n\underline{\underline{e}}_{i,ni} ;\end{cases}$$

1.12
$$\kappa(\Sigma_{m,n} V_m[x_{m,n}]F_n) = \Sigma_{m,n,i\in\underline{\underline{P}}}\ n x^i_{m,n}\ \underline{\underline{e}}_{mi,ni}$$

Indeed, formulas (1.11) express only the computations in (1.8), while (1.12) follows immediately from (1.11). A direct inspection of (1.12) shows that

1.13 <u>the ring homomorphism</u> $\kappa : \text{Cart}(K) \to M_{(\underline{\underline{P}})}(K)$ <u>is injective iff</u> K <u>is torsion free</u>.

Due to (1.5), we need only check formulas (1.7) when $K = \underline{\underline{Z}}$ or $K = \underline{\underline{Z}}[c]$, a polynomial ring. But then K is injective (1.13), so that it suffices to check the equality of both sides in each of the formulas (1.7) <u>after applying</u> κ. Then they follow immediately from (1.11).

2. <u>An axiomatic description of the functor</u> Cart_S

2.1 Let S be any <u>set of primes</u>. We denote by $\underline{\underline{S}}$ the set of the integers, all the prime factors of which belong to S, in other words the multiplicative closure of S. Note that, if $S = \emptyset$, then $\underline{\underline{S}} = \{1\}$.

2.2 <u>The functor</u> Cart_S <u>associates to a basic ring</u> K <u>a topological (generally non commutative) ring</u> $\text{Cart}_S(K)$, <u>which is (topologically) generated by elements</u> V_n, F_n, <u>where</u> $n \in \underline{\underline{S}}$, <u>and</u> $[c]$, <u>where</u> $c \in K$.

<u>Every element</u> $x \in \text{Cart}_S(K)$ <u>admits a unique representation as a converging sum</u>,

2.3
$$x = \Sigma_{m,n\in\underline{\underline{S}}}\ V_m[x_{m,n}]F_n ,$$

with the finiteness condition

2.4 $$(x_{m,n})_{n\in \underline{\underline{S}}} \in K^{(\underline{\underline{S}})} \quad \text{for every} \quad m \in \underline{\underline{S}} .$$

For any homomorphism of basic rings, $\varphi : K \to K'$, the ring homomorphism $\varphi_* : \mathrm{Cart}_S(K) \to \mathrm{Cart}_S(K')$ maps $x \in \mathrm{Cart}_S(K)$, written as in (2.3), on

2.5 $$\varphi_* x = \Sigma_{m,n\in \underline{\underline{S}}} V_m[\varphi(x_{m,n})]F_n \in \mathrm{Cart}_S(K') .$$

The order of $x \in \mathrm{Cart}_S(K)$, written as (2.3), is the least $m \in \underline{\underline{S}}$ for which there is an $x_{m,n} \neq 0$ (if $x_{m,n} = 0$ for all m,n, i.e. if $x = 0$, we define the order of x as $+\infty$). The order function satisfies the following axioms:

2.6 $$\mathrm{ord}(x \pm y) \geq \min(\mathrm{ord}(x), \mathrm{ord}(y)) ,$$

2.7 $$\mathrm{ord}(xy) \geq \mathrm{ord}(x) .$$

The topology of $\mathrm{Cart}_S(K)$ is defined by its order function, and $\mathrm{Cart}_S(K)$ is complete.

The unit 1 of $\mathrm{Cart}_S(K)$ is

2.8 $$1_{\mathrm{Cart}_S(K)} = [1_K] = V_1 = F_1 .$$

For any $m,n \in \underline{\underline{S}}$ and $a,b \in K$, we have

2.9 $$V_m V_n = V_{mn} , \; F_m F_n = F_{mn} , \; [a][b] = [ab] ;$$

2.10 $$[a]V_m = V_m[a^m] , \; F_n[a] = [a^n]F ;$$

2.11 $$F_n V_n = n\, 1_{\mathrm{Cart}_S(K)} .$$

2.12 If $\gcd(m,n) = 1$, then $V_m F_n = F_n V_m$;

2.13 $$\mathrm{ord}([a+b] - [a] - [b]) > 1 .$$

We shall prove in the next section that the above axioms actually define a functor Cart_S .

2.14 Qua functor in sets, $\mathrm{Cart}_S(K)$ may be identified with $M_{(\underline{S})}(K)$ (see 1.9): this is expressed by (2.3), (2.4), (2.5).

In $\mathrm{Cart}_S(K)$, the left multiplication by V_n and the right multiplication by F_n are injective, and

2.15 $$\mathrm{ord}(V_n x) = n\ \mathrm{ord}(x)\ ,\ \mathrm{ord}(xF_n) = \mathrm{ord}(x)\ ,$$

for any $x \in \mathrm{Cart}_S(K)$: this follows from (2.3), (2.9).

2.16 Lemma. For any $x,y \in K$ and $m,n,i,j \in \underline{S}$, we have

2.17 $$V_m[x]F_n\ V_i[y]F_j = dV_{mi/d}[x^{i/d}y^{n/d}]F_{nj/d}\ ,$$

where $d = \gcd(n,i)$.

Proof. We apply (2.9) to write

$$F_n\ V_i = F_{n/d}\ F_d\ V_d\ V_{i/d}\ ,$$

and (2.11), (2.12) to obtain

$$F_n\ V_i = d\ V_{i/d}\ F_{n/d}\ .$$

Then axiom (2.10) gives

$$[x]F_n V_i[y] = d\ V_{i/d}[x^{i/d}][y^{n/d}]F_{n/d}\ ,$$

and formula (2.17) follows from (2.9).

2.18 For any $n \in \underline{S}$ and $x \in \mathrm{Cart}_S(K)$, we have

$$\mathrm{ord}(F_n x) \geq n^{-1}\mathrm{ord}(x)\ .$$

This formula, analogous to (III.3.27), follows from lemma (2.16) and the properties of the order function: see (2.3), (2.6), (2.7). It is used to check that $\mathrm{Cart}_S(K)$ is actually a topological ring: see (III.7).

3. Properties and existence of Cart_S are derived from the matrix representation κ'

3.1 For brevity's sake, let us write $E(K)$, or simply E, instead of $\mathrm{Cart}_S(K)$.

3.2 For any $i \in \underline{\underline{S}}$, we denote by E_i (resp. E_{i+}) the subset of E defined by the condition $\mathrm{ord}(x) \geq i$ (resp. $\mathrm{ord}(x) > i$). If x is written as in (2.3), then $x \in E_i$ (resp. $x \in E_{i+}$) means that $x_{m,n} = 0$ when $m < i$ (resp. $m \leq i$).

It follows from (2.6), (2.7), that E_i and E_{i+} are <u>right ideals</u> in E. We put

3.3 $$\mathrm{gr}_i(E) = E_i/E_{i+},$$

and we obtain, for any $i \in \underline{\underline{S}}$, a <u>right</u> E-<u>module</u> $\mathrm{gr}_i(E)$. As V_i induces an E-<u>module isomorphism</u>, $\mathrm{gr}_1(E) \to \mathrm{gr}_i(E)$, for every i, it will suffice to study $\mathrm{gr}_1(E)$.

3.4 <u>There is a natural structure of free</u> K-<u>module on</u> $\mathrm{gr}_1(E)$, <u>with basis</u> $(\bar{F}_i)_{i \in \underline{\underline{S}}}$, <u>where</u> $\bar{F}_i$ <u>is</u> F_i <u>modulo</u> E_{1+}. <u>By definition</u>, $\Sigma_{i \in \underline{\underline{S}}}\, a_i \bar{F}_i = \Sigma_{i \in \underline{\underline{S}}}\, [a_i] F_i$ <u>modulo</u> E_{1+}, $(a_i)_{i \in \underline{\underline{S}}} \in K^{(\underline{\underline{S}})}$.

3.5 Indeed, axioms (2.13) and (2.6) imply that

$$\Sigma_{i \in \underline{\underline{S}}}[a_i]F_i + \Sigma_{i \in \underline{\underline{S}}}[b_i]F_i \equiv \Sigma_{i \in \underline{\underline{S}}}[a_i + b_i]F_i \quad \underline{\text{modulo}}\ E_{1+},$$

while it follows from (2.3) that the elements

$$\Sigma_{i \in \underline{\underline{S}}}\, [a_i]F_i \ , \quad (a_i)_{i \in \underline{\underline{S}}} \in K^{(\underline{\underline{S}})}$$

are <u>a full set of representatives of</u> E_1 <u>modulo</u> E_{1+}.

3.6 As <u>the ring of all</u> K-<u>linear endomorphisms of the free</u> module $K^{(\underline{\underline{S}})}$, <u>acting on the right, may be identified with the matrix ring</u> $M_{(\underline{\underline{S}})}(K)$ (see 1.9), acting on the right, we can express the right action of E on $\mathrm{gr}_1(E)$ by a ring homomorphism $\kappa' : E(K) \to M_{(\underline{\underline{S}})}(K)$.

The following equalities in $\mathrm{gr}_1(E)$:

3.7 $$\bar{F}_i \cdot F_n = \overline{F_{in}}\ ;$$

3.8 $$\bar{F}_i \cdot V_n = \begin{cases} 0 & \text{if } n \nmid i , \\ n\bar{F}_{i/n} & \text{if } n \mid i , \end{cases}$$

3.9 $$\bar{F}_i \cdot [c] = c^i \bar{F}_i \quad .$$

result from axioms (2.9), (2.10) and lemma (2.16). They are equivalent to the following matrix equalities in $M_{(\underline{\underline{S}})}(K)$:

3.10 $$\begin{aligned} \kappa'(F_n) &= \Sigma_{i \in \underline{\underline{S}}} \; \underline{\underline{e}}_{i,ni} \quad , \\ \kappa'(V_n) &= \Sigma_{i \in \underline{\underline{S}}} \; n\underline{\underline{e}}_{ni,i} \quad , \\ \kappa'([c]) &= \Sigma_{i \in \underline{\underline{S}}} \; c^i e_{i,i} \quad . \end{aligned}$$

3.11 <u>There is a functorial ring homomorphism,</u>

$$\kappa' \; : \; \mathrm{Cart}_S(K) \to M_{(\underline{\underline{S}})}(K) \quad ,$$

<u>which maps</u> $x = \Sigma_{m,n \in \underline{\underline{S}}} V_m [x_{m,n}] F_n \in \mathrm{Cart}_S(K) \; (=E)$ on

3.12 $$\kappa'(x) = \Sigma_{m,n,i \in \underline{\underline{S}}} \; m x^i_{m,n} \, \underline{\underline{e}}_{mi,ni} \in M_{(\underline{\underline{S}})}(K) \quad .$$

3.13 <u>Remark</u>. The ring homomorphism κ', given by (3.12), does not reduce to κ when $S = P$ (see 1.12). Indeed, there is for any S a functorial ring homomorphism $\kappa : \mathrm{Cart}_S(K) \to M_{(\underline{\underline{S}})}(K)$, which maps $\Sigma_{m,n \in \underline{\underline{S}}} V_m [x_{m,n}] F_n$ on

3.14 $$\kappa(x) = \Sigma_{m,n,i \in \underline{\underline{S}}} \; n x^i_{m,n} \, \underline{\underline{e}}_{mi,ni} \quad .$$

Let us denote by N the diagonal matrix $\Sigma_{i \in \underline{\underline{S}}} \; i \, \underline{\underline{e}}_{i,i}$. Then

3.15 $$N\kappa(x) = \kappa'(x) N \quad .$$

While κ gives the action of E on power series $\Sigma_{i \in \underline{\underline{S}}} a_i t^i$, κ' gives the action of E on their <u>derivatives</u>.

The following considerations could be derived as well from κ as from κ' .

In order to compute the matrix entry $\kappa'(x)_{m,n}$ in (3.12), i.e. to sum the coefficients of $\underline{\underline{e}}_{m,n}$, we have to introduce

$r = \gcd(m,n)$, $\mu = mr^{-1}$, $\nu = nr^{-1}$. Then we obtain:

3.16 $$\kappa'(x)_{\mu r,\nu r} = \mu \sum_{d|r} d\, x^{r/d}_{\mu d,\nu d}\ , \text{ where } \mu,\nu,r \in \underline{\underline{S}}\ ,\ \gcd(\mu,\nu) = 1\ .$$

3.17 In order to draw the consequences of (3.16), let us say that a "monomial" $V_m[x_{m,n}]F_n$ has the <u>slope</u> m/n, and let us denote by Sl S the set of all possible slopes, i.e. the multiplicative group $\underline{\underline{S}}\,\underline{\underline{S}}^{-1}$ generated by S in $\underline{\underline{Q}}$. For any $\varrho \in$ Sl S, let E_ϱ be the subset of E consisting of the elements $\Sigma_{m,n\in\underline{\underline{S}},m=\varrho n} V_m[x_{m,n}]F_n$; in other words, all monomials in the expansion of x (2.3) must have the same slope ϱ.

3.18 <u>Theorem</u>. <u>If</u> $\varrho = \mu/\nu$, <u>where</u> $\mu,\nu \in \underline{\underline{S}}$, $\gcd(\mu,\nu) = 1$, <u>then</u>

3.19 $$E_\varrho = V_\mu\, E_1\, F_\nu\ ,$$

<u>and</u> E_ϱ <u>is a closed additive subgroup of</u> E. <u>Every</u> $x \in E$ <u>has a unique expansion as</u>

3.20 $$x = \Sigma_{\varrho\in \mathrm{Sl}\ S}\, x_\varrho\ ,$$

<u>where</u> $x_\varrho \in E_\varrho$ <u>and</u> $(x_\varrho)_{\varrho\in \mathrm{Sl}\ S}$ <u>converges towards</u> 0 <u>in</u> E. <u>For any</u> $\varrho,\sigma \in$ Sl S, <u>we have</u>

3.21 $$E_\varrho E_\sigma \subset E_{\varrho\sigma}\ ,$$

<u>and</u> E_1 <u>is a commutative subring of</u> E.

Proof. Formulas (3.19) and (3.20) are obtained simply by collecting "monomials" of a given slope in the general expansion (2.3) of an $x \in E$.

If K is torsion free, than formula (3.16) imply that κ' is injective, and $x \in E_1$ means that $\kappa'(x)$ is a <u>diagonal matrix</u>. Therefore E_1 is a <u>commutative subring</u> of E, and E_ϱ is a closed additive subgroup for any $\varrho \in$ Sl S. As Cart_S is a <u>functor</u>, all this still holds for any basic ring K.

Finally, lemma (2.16) shows that the product of any two "monomials" of slopes ϱ,σ is an integral multiple of a "monomial" of slope $\varrho\sigma$. By the distributivity axiom of rings, this proves (3.21).

3.22 Remark. We could say that E is a topological graded ring, by saying "degree" instead of slope, and specifying that "degrees of homogenous elements in E are multiplied (not added!) when such elements are multiplied".

3.23 Corollary to theorem (3.18). Let $\underline{\underline{A}}$ be a subset of Sl S, closed for multiplication and containing 1 (for instance, the set of the slopes $\varrho \geq 1$, or ≤ 1). Then the subset $E_{\underline{\underline{A}}}$ of E, made up of all converging sums $\Sigma_{\varrho\in\underline{\underline{A}}} x_\varrho$, $x_\varrho \in E_\varrho$, is a subring of E.

3.24 Theorem. The axioms (2.2) to (2.13) are categorical, i.e. there is one functor Cart_S completely defined by these axioms.

Proof. The axioms define E(K) qua functor in sets (2.14), and the matrix representation κ' (3.11) provides universal formulas for computing sums and products in E(K). But, a priori, these formulas are polynomial over $\underline{\underline{Q}}$, not over $\underline{\underline{Z}}$, and the point is to check that they are indeed defined over $\underline{\underline{Z}}$. Now, for the additive subgroup E_1, this was done in (III.2.15), since

3.25 $$\kappa'(\Sigma_{n\in\underline{\underline{S}}} V_n[x_n]F_n) = \Sigma_{n\in\underline{\underline{S}}} (\Sigma_{d|n} dx_d^{n/d})\underline{\underline{e}}_{n,n} .$$

It follows from (3.19) and (3.20) that addition is defined in E(K) by polynomial formulas with integral coefficients. As concernes multiplication of "monomials", lemma (2.16) shows that their products are integral multiple of "monomials", and therefore are computable through universal formulas with integral coefficients. By the distributivity law, this holds for all products. So we have proved the existence of a "non standard" ring structure on the underlying set $M_{(\underline{\underline{S}})}(K)$, for which the axioms (2.2) to (2.13) are easily checked.

4. The ring $W_S(K)$ and its integers

4.1 Definition. Let K be a basic ring and S a set of primes (see 2.1). Then we denote by $W_S(K)$ the set $K^{\underline{\underline{S}}}$ endowed with the topological commutative ring structure defined by the injective ring homomorphism $op : W_S(K) \to Cart_S(K)$, as follows:

4.2 $$op(x) = \Sigma_{n \in \underline{\underline{S}}} V_n[x_n]F_n \text{ , where } x = (x_n)_{n \in \underline{\underline{S}}} \in W_S(K) .$$

4.3 In other words, $W_S(K)$ is isomorphic to the commutative ring E_1 of theorem (3.18). The reason why we do not define $W_S(K)$ as a subring of $Cart_S(K)$ is that there will be a natural left action of $Cart_S(K)$ on $W_S(K)^+$ (i.e. $W_S(K)$ qua additive group), so that identifying $W_S(K)$ with its canomical image in $Cart_S(K)$ would lead to confusion (see 4.19).

4.4 The topology of $W_S(K)$ is the product topology of $K^{\underline{\underline{S}}}$, where K is discrete. The ring structure is defined by the functorial ring homomorphisms $w_n : W_S(K) \to K$, $n \in \underline{\underline{S}}$:

4.5 $$w_n(x) = \Sigma_{d|n} dx_d^{n/d} \text{ , for } x = (x_n)_{n \in \underline{\underline{S}}} \in W_S(K) .$$

4.6 The formulas giving the sum $z = x + y$ and the product $t = xy$ of any two elements $x,y \in W_S(K)$ are polynomial, with integral coefficients. More precisely, z_n and t_n are polynomials with integral coefficients in the x_d and y_d, where $d|n$. If each x_d,y_d receives the weight d, then z_n is an isobaric polynomial of weight n; if x_d receives the weight

$(d,0) \in \underline{\underline{N}}^2$ and y_d the weight $(0,d)$, then t_n is an isobaric polynomial of weight (n,n), or, with the preceding weights, t_n is isobaric of weight n separately in n and in y (and consequently of total weight $2n$). All this follows from the relations

4.7 $$w_n(z) = w_n(x) + w_n(y) \; ; \quad w_n(t) = w_n(x)w_n(y) \quad , n \in \underline{\underline{S}} .$$

4.8 The ring $W_S(K)$ appears as a <u>factor ring</u> of the ring $W(K) = W_P(K)$, which we call the <u>ring of Witt vectors over</u> $\underline{\underline{Z}}$, <u>with coordinates in</u> K. The formulas for sums and products in $W_S(K)$ are the same as in $W(K)$, coordinates with index $n \notin \underline{\underline{S}}$ beeing disregarded.

4.9 The unit of $W_S(K)$ is the vector $(1,0,0,\ldots)$, i.e. $(x_n)_{n\in\underline{\underline{S}}}$, with $x_1 = 1$, $x_n = 0$ for $n > 1$ (see 2.8) In order to compute the integers in $W_S(K)$, i.e. the integral multiples of the unit, we may apply the homomorphisms w_n.

4.10 <u>For any</u> $k \in \underline{\underline{Z}}$, <u>we have in</u> $W_S(K)$

$$k \cdot 1_{W_S(K)} = (u_{k,n} \cdot 1_K)_{n\in\underline{\underline{S}}} ,$$

<u>where the integers</u> $u_{k,n} \in \underline{\underline{Z}}$ <u>are defined by the relations</u>

4.11 $$\Sigma_{d|n} \; d u_{k,d}^{n/d} = k \quad , \quad \text{for any } n \in \underline{\underline{S}} .$$

Remembering how the additive group $W(K)^+$ was defined by its isomorphism with $\mathfrak{C}(\underline{\underline{G}}_a)$ (see III.1.11), we see that, for any $k \in \underline{\underline{Z}}$, the <u>complete</u> sequence $(u_{k,n})_{n\in\underline{\underline{P}}}$ is defined by the following equality of formal series in one letter t :

4.12 $$(1-t)^k = \Pi_{n\in\underline{\underline{P}}} (1-u_{k,n}t^n) .$$

4.13 <u>Proposition</u>. <u>An integer</u> $k \in \underline{\underline{Z}}$ <u>is invertible in</u> $W_S(K)$ <u>and in</u> $\mathrm{Cart}_S(K)$ <u>iff it is invertible in</u> K.

Proof. If k is invertible in $W_S(K)$, or equivalently in $Cart_S(K)$, according to theorem (3.18), then we see that k is invertible in K by applying the ring homomorphism w_1.

Now, the binomial coefficients $\binom{k^{-1}}{n}$, $n \in \underline{\underline{N}}$, lie in the ring $\underline{\underline{Z}}[k^{-1}]$. Here is an application of the isomorphism theorem (I.8.1). The morphism $t \mapsto 1-(1-t)^k$ is an isomorphism over $\underline{\underline{Z}}[k^{-1}]$ because the coefficient of t, i.e. k, is invertible in this ring. The inverse morphism is

$$t \mapsto 1-(1-t)^{k^{-1}} = \Sigma_{n \in \underline{\underline{P}}} (-1)^{n-1} \binom{k^{-1}}{n} t^n .$$

Therefore, if k is invertible in K, it is also invertible in $W(K)$, the coordinates $u_{k^{-1},n}$ of its inverse beeing defined by the equality of formal series

4.14 $$(1-t)^{k^{-1}} = \Pi_{n \in \underline{\underline{P}}} (1-u_{k^{-1},n} t^n) .$$

4.15 Lemma. Let p be a prime. Then

$$u_{p,n} \equiv 0 \text{ mod. } p \text{ for any } n \neq p , \text{ and}$$
$$u_{p,p} \equiv 1 \text{ mod. } p .$$

Proof. We put $k = p$ in (4.12) and we reduce mod.p. As $(1-t)^p \equiv 1-t^p$ mod. p and the factorization in the right side is unique, we obtain the above relations.

4.16 Proposition. Let $p \in S$. Then the following relations are equivalent:

4.17 $$p\,1_K = 0 \text{ in } K ;$$

4.18 $$F_p V_p = V_p F_p \text{ in } W_S(K) .$$

Indeed, (4.17) implies (4.18) according to lemma (4.15) (see 2.11). Conversely, (4.18) implies (4.17), because $u_{k,1} = k$ for any $k \in \underline{\underline{Z}}$ (see 4.12).

4.19 Proposition. For every $m \in \underline{\underline{S}}$, there is a functorial ring endomorphism of $W_S(K)$, denoted by $x \mapsto F_m \cdot x$, which satisfies the following relations in $\mathrm{Cart}_S(K)$

4.20 $$F_m \mathrm{op}(x) = \mathrm{op}(F_m \cdot x) F_m \quad , \qquad x \in W_S(K) ,$$

4.21 $$\mathrm{op}(x) V_m = V_m \, \mathrm{op}(F_m \cdot x) \quad , \qquad x \in W_S(K) ,$$

and is defined by either of them. For any $n \in \underline{\underline{S}}$, we have

4.22 $$w_n(F_m \cdot x) = w_{nm}(x) \quad , \quad x \in W_S(K) .$$

Proof. With the notations of theorem (3.18), the elements $F_m \, \mathrm{op}(x)$ and $\mathrm{op}(x) V_m$ lie respectively in $E_{m^{-1}}$ and E_m , which insures the existence and unicity of elements $x', x'' \in W_S(K)$ such that

4.23 $$F_m \, \mathrm{op}(x) = \mathrm{op}(x') F_m \quad ,$$

4.24 $$\mathrm{op}(x) V_m = V_m \, \mathrm{op}(x'') .$$

From these relations, we deduce that the maps $x \mapsto x'$ and $x \mapsto x''$ are endomorphisms of $W_S(K)$. As $\mathrm{op}(x)$ commutes with $V_m F_m$, we have

$$V_m \mathrm{op}(x') F_m = V_m F_m \mathrm{op}(x) = \mathrm{op}(x) V_m F_m = V_m \mathrm{op}(x'') F_m ,$$

whence $x' = x''$, and we define $F_m \cdot x = x' = x''$.

In order to prove (4.22), we may use the matrix representation κ' (3.11), related to $W_S(K)$ by the formula

4.25 $$\kappa'(\mathrm{op}(x)) = \Sigma_{i \in \underline{\underline{S}}} \, w_i(x) \underline{\underline{e}}_{i,i} .$$

When applying κ' to both sides of (4.23), we have, according to (3.10).

$$\Sigma_{i \in \underline{\underline{S}}} \, w_{mi}(x) \underline{\underline{e}}_{i,mi} = \Sigma_{i \in \underline{\underline{S}}} \, w_i(F_m \cdot x) \underline{\underline{e}}_{i,mi} ,$$

which proves (4.22).

4.26 *Corollary: a generalization of lemma* (2.16). *For any* $x, y \in W_S(K)$ *and* $m, n, i, j \in \underline{\underline{S}}$, *we have*

$$V_m \mathrm{op}(x) F_n \; V_i \mathrm{op}(y) F_j = V_{mi/d} \; \mathrm{op}(d(F_{i/d} \cdot x)(F_{n/d} \cdot y)) F_{nj/d} ,$$

where $d = \gcd(n,i)$.

The proof is that of lemma (2.16), replacing (2.10) by (4.20) and (4.21), and using the fact that "op" is a ring homomorphism to place the integral factor d "inside", not "outside".

5. *Uniform and reduced* $\mathrm{Cart}_S(K)$-*modules*

5.1 The present section is mainly a repetition of section (III.7). Here we consider a fixed basic ring K, and a fixed set of primes S. We may assume that $S \neq P$.

5.2 *Definition. A uniform* $\mathrm{Cart}_S(K)$-*module is a left topological* $\mathrm{Cart}_S(K)$-*module* C , *where all sums* $\Sigma_{j \in J}\, x_j \cdot \gamma_j$, $x_j \in \mathrm{Cart}_S(K)$, $\gamma_j \in C$, *converge provided that* $(x_j)_{j \in J}$ *converges towards* 0 *in* $\mathrm{Cart}_S(K)$.

5.3 In a uniform $\mathrm{Cart}_S(K)$-module C , the additive subgroup C_n is defined only for $n \in \underline{\underline{S}}$ (not $n \in \underline{\underline{P}}$) as the closure of the sum of all subgroups $V_i \cdot C$ for $i \in \underline{\underline{S}}$, $i \geq n$.

5.4 Proposition (III.7.7) holds for uniform $\mathrm{Cart}_S(K)$-modules, with Cart(K) replaced by $\mathrm{Cart}_S(K)$ both in the statement and in its proof (otherwise unchanged). *So* C *is filtered complete for the* (C_n)-*topology, which is finer than its given topology*.

5.5 Sometimes it will be convenient to use the *order function* on an uniform $\mathrm{Cart}_S(K)$-module C , i.e. the function $\mathrm{ord} : C \to \underline{\underline{S}} \cup \{+\infty\}$, defined by the relations

5.6 $$\mathrm{ord}(\gamma) \geq m \quad \text{iff} \quad \gamma \in C_m$$

for any $\gamma \in C$ and any $m \in \underline{\underline{S}}$.

5.7 The category of uniform $\mathrm{Cart}_S(K)$-modules is defined by taking as morphisms the $\mathrm{Cart}_S(K)$-linear continuous maps.

Instead of defining $gr_n C$ as the factor group C_n/C_{n+1} (see III.7.13), we put

5.8 $$gr_n C = C_n/C_{n+} \quad ,$$

where (here!) $n+$ denotes the successor of n in $\underline{\underline{S}}$ (not in $\underline{\underline{P}}$, generally). Then we have additive maps $gr_n C \to gr_{mn} C$ induced by V_m , and $gr_1 C$ has a natural structure of K-module.

5.9 Definition. A uniform $\mathrm{Cart}_S(K)$-module is said to be reduced iff

5.10 its topology is the order topology;

5.11 the map $gr_1 C \to gr_m C$ induced by V_m is bijective for any $m \in \underline{\underline{S}}$;

5.12 the K-module $gr_1 C$ is free.

As the map $gr_1 C \to gr_m C$ is always surjective, it suffices to require that it be injective. The two conditions (5.11) and (5.12) are equivalent to the existence of a V-basis in C (see III.11.2).

5.13 Definition. A V-basis in a uniform $\mathrm{Cart}_S(K)$-module C is an indexed set of elements $(\gamma_i)_{i \in I}$, which verifies the two following equivalent conditions:

5.14 for any $n \in \underline{\underline{S}}$, the elements $\Sigma_{i \in I} V_n[x_i] \cdot \gamma_i$, where $(x_i)_{i \in I} \in K^{(I)}$, are a full set of representatives of $gr_n C$ in C_n ;

5.15 every $\gamma \in C$ has a unique expansion as

$$\gamma = \Sigma_{i\in I, n\in \underline{\underline{S}}} V_n[x_{n,i}]\cdot\gamma_i ,$$

where $(x_{n,i})_{i\in I} \in K^{(I)}$ for any n $\underline{\underline{S}}$.

5.16 The <u>category of reduced</u> $\mathrm{Cart}_S(K)$-modules is defined as a full subcategory of that of the uniform $\mathrm{Cart}_S(K)$-modules (see 5.7).

5.17 <u>If</u> C <u>is a reduced</u> $\mathrm{Cart}_S(K)$-<u>module, than an indexed set</u> $(\gamma_i)_{i\in I}$ <u>in</u> C <u>is a</u> V-<u>basis iff the natural images of the</u> γ_i <u>in</u> $\mathrm{gr}_1 C$ <u>are a basis of this free</u> K-<u>module</u>.

5.18 The category of uniform $\mathrm{Cart}_S(K)$-modules is much larger than that of reduced $\mathrm{Cart}_S(K)$-modules (if $S \neq \emptyset$). Indeed, if C is a uniform $\mathrm{Cart}_S(K)$-module, so is every closed submodule C' of C , and every factor module C/C' (with closed kernel C').

The order function on $C' \subset C$ is not (in general) the restriction to C' of the order function of C . But the order function on C/C' is the "factor order function", i.e. C_n maps <u>onto</u> $(C/C')_n$ for any $n \in \underline{\underline{S}}$. In other words, if $f : C \to C/C'$ denotes the natural map, then, for any $\delta \in C/C'$,

5.19 $$\mathrm{ord}(\delta) = \max \mathrm{ord}(\gamma) \quad , \quad \text{for} \quad \gamma \in f^{-1}(\delta) .$$

The maximum is reached in (5.19) because C' is assumed to be closed.

6. <u>The functorial ring homomorphism</u> $\kappa_{T,S}$

6.1 Let T be a set of primes, S a proper subset of T, and $U = T-S$ its complement. In order to avoid confusion when considering elements in the rings $\mathrm{Cart}_T(K)$ and $\mathrm{Cart}_S(K)$, we shall (exceptionnally) use superscripts to mention the relevant set of primes, writing $V_n^T \in \mathrm{Cart}_T(K)$, etc. Note that any $n \in \underline{\underline{T}}$ has a unique representation as $n = i\alpha$, with $i \in \underline{\underline{S}}$, $\alpha \in \underline{\underline{U}}$. We put $E_T = \mathrm{Cart}_T(K)$, $E_S = \mathrm{Cart}_S(K)$.

6.2 Lemma. Let $E_{T,S}$ be the set of elements $x \in E_T$ which are written as

6.3 $$x = \Sigma_{i,j\in\underline{\underline{S}},\alpha\in\underline{\underline{U}}} V_{i\alpha}^T[x_{i,j,\alpha}]^T F_{j\alpha}^T ,$$

where $(x_{i,j,\alpha})_{j\in\underline{\underline{S}}} \in K^{(\underline{\underline{S}})}$ for any $i \in \underline{\underline{S}}$, $\alpha \in \underline{\underline{U}}$ (see 2.3). Further, let $I_{T,S} \subset E_{T,S}$ be defined by the condition $x_{i,j,1} = 0$ for any $i,j \in \underline{\underline{S}}$ imposed on x in (6.3). Then $E_{T,S}$ is a subring of E_T, and $I_{T,S}$ is the kernel of a ring homomorphism onto E_S, mapping $x \in E_{T,S}$ in (6.3) on

6.4 $$\Sigma_{i,j\in\underline{\underline{S}}} V_i^S[x_{i,j,1}]^S F_j^S \in E_S .$$

6.5 Proof. The statement that $E_{T,S}$ is a subring of E_T is only a special case of (3.23), by putting $\underline{\underline{A}} = \mathrm{Sl}\ S$. The subset $I_{T,S}$ is closed for addition, because the "coordinate" $z_{i,j,1}$, in a sum $z = x + y$ of elements of $I_{T,S}$, depends only on the "coordinates" $x_{i',j',1}$, $y_{i',j',1}$ of x and y such that $i' = i/d$, $j' = j/d$ for some integer d, and therefore $z_{i,j,1} = 0$ (see 3.19 and 4.6). If $x = V_{i\alpha}^T[a]^T F_{j\alpha}^T$ is a "monomial" in $I_{T,S}$ then, when x is multiplied on the left by any $y \in E_{T,S}$, it keeps its right factor $F_\alpha^T(\alpha\in\underline{\underline{U}}, \alpha>1)$, and therefore $yx \in I_{T,S}$; similarly $xy \in I_{T,S}$ and $I_{T,S}$ is a two-sided ideal.

Now the set of elements $\Sigma_{i,j\in\underline{\underline{S}}} V_i^T[x_{i,j}]^T F_j^T$ is a full set of representatives of $E_{T,S}$ modulo $I_{T,S}$. All axioms for elements of E_S (see 2.2 to 2.13) are verified by those representatives modulo $I_{T,S}$, and as the axioms are categorical (3.24), our lemma is proved.

6.6 *Theorem. Let* $\tilde{M}_{\underline{\underline{U}}}(E_S)$ *denote the ring of infinite square matrices* $(x_{\alpha,\beta})_{\alpha,\beta\in\underline{\underline{U}}}$ *with entries from* E_S, *and such that every row* $(x_{\alpha,\beta})_{\beta\in\underline{\underline{U}}}$ *converges towards* 0. *Then there is a functorial ring homomorphism,*

$$\kappa_{T,S} : E_T \to \tilde{M}_{\underline{\underline{U}}}(E_S) ,$$

defined by the following formulas using matrix units (where $i,j \in \underline{\underline{S}}$, $\alpha,\beta \in \underline{\underline{U}}$, $c \in K$).

6.7 $$\kappa_{T,S}(V^T_{i\alpha}) = \Sigma_{\lambda\in\underline{\underline{U}}} V^S_i \underline{\underline{e}}_{\lambda\alpha,\lambda} ,$$

6.8 $$\kappa_{T,S}([c]^T) = \Sigma_{\lambda\in\underline{\underline{U}}}[c^\lambda]^S \underline{\underline{e}}_{\lambda,\lambda} ,$$

6.9 $$\kappa_{T,S}(F^T_{j\beta}) = \Sigma_{\lambda\in\underline{\underline{U}}} \beta F^S_j \underline{\underline{e}}_{\lambda,\lambda\beta} , \quad \text{and}$$

6.10 $$\kappa_{T,S}(\Sigma_{i,j\in\underline{\underline{S}},\alpha,\beta\in\underline{\underline{U}}} V^T_{i\alpha}[x_{i,j,\alpha,\beta}]^T F^T_{j\beta}) =$$
$$= \Sigma_{i,j\in\underline{\underline{S}},\alpha,\beta,\lambda\in\underline{\underline{U}}} \beta V^S_i [x^\lambda_{i,j,\alpha,\beta}]^S F^S_j \underline{\underline{e}}_{\lambda\alpha,\lambda\beta}$$

6.11 Proof. Let R be the closed left ideal generated in E_T by the elements F_p , $p \in U$ and $C = E_T/R$ the factor uniform E_T-module (see 5.18). The right multiplication by some $x \in E_T$ (i.e. $y \mapsto yx$) induces an *endomorphism of* C *iff* $Rx \subset R$: it is so for any $x \in E_{T,S}$. Indeed, it follows from (3.18) that $F_pE_{T,S} \subset E_{T,S}F_p$ for any $p \in U$.

So $E_{T,S}$ acts on the right of the left E_T-module C . An element $x \in E_{T,S}$ acts as 0 iff $x \in E_{T,S} \cap R$, and $E_{T,S} \cap R = I_{T,S}$. Then we can consider C as a right $(E_{T,S}/I_{T,S})$-module, i.e. as a right E_S-module by using the natural isomorphism $E_{T,S} \to E_S$ of lemma (6.2).

Now we obtain a full set of representatives of E_T modulo R by taking the elements

6.12 $$\Sigma_{i,j\in\underline{\underline{S}},\alpha\in\underline{\underline{U}}} V^T_{i\alpha}[x_{i,j,\alpha}]^T F^T_j .$$

6.13 Let us denote by $\bar{V}_\alpha$ the natural image of V_α^T in C, $\alpha \in \underline{\underline{U}}$. Formula (6.12) says that any $\xi \in C$ can be written as $\xi = \Sigma_{\alpha \in \underline{\underline{U}}} \bar{V}_\alpha x_\alpha$, with well defined $x_\alpha \in E_S$, subject to no condition. Replacing $\xi \in C$ by $(x_\alpha)_{\alpha \in \underline{\underline{U}}} \in E_S^{\underline{\underline{U}}}$, we may identify C, qua topological right E_S-module, with the product $E_S^{\underline{\underline{U}}}$.

6.14 Any continuous E_S-linear endomorphism $f : C \to C$ maps $\bar{V}_\beta$ on $f(\bar{V}_\beta) = \Sigma_{\alpha \in \underline{\underline{U}}} \bar{V}_\alpha u_{\alpha,\beta}$, $u_{\alpha,\beta} \in E_S$, and, as $f(\bar{V}_\beta)$ converges towards 0, $(u_{\alpha,\beta})_{\beta \in \underline{\underline{U}}}$ has to converge towards 0 for any $\alpha \in \underline{\underline{U}}$, i.e. with our notation $(u_{\alpha,\beta})_{\alpha,\beta \in \underline{\underline{U}}} \in \tilde{M}_{\underline{\underline{U}}}(E_S)$. Conversely, any such matrix defines f by the formula $f(\Sigma_{\alpha \in \underline{\underline{U}}} \bar{V}_\alpha x_\alpha) = \Sigma_{\alpha \in \underline{\underline{U}}} \bar{V}_\alpha (\Sigma_{\beta \in \underline{\underline{U}}} u_{\alpha,\beta} x_\beta)$. Taking for f the left action of $x \in E_T$ on C, we obtain the matrix $\kappa_{T,S}(x)$. Formulas (6.7), (6.8), (6.9) are checked in the standard way, and (6.10) follows from them.

6.15 Remark. If $T = P$ and $S = \emptyset$, then $E_S = K$, $E_T = \mathrm{Cart}(K)$, and $\kappa_{P,\emptyset}$ is the functorial homomorphism κ of (1.10).

7. The category of S-typical groups

7.1 Let S be a fixed set of primes, $U = P - S$ its complement and K a basic ring. We say that *an element* γ *in a left* $\mathrm{Cart}(K)$*-module* B *is* S*-typical iff*

7.2 $$F_p \cdot \gamma = 0 \quad \text{for any } p \in U \quad (\text{i.e. } p \notin S).$$

7.3 *Lemma. Let* B_S *be the set of all* S*-typical elements in a left* $\mathrm{Cart}(K)$*-module* B*. Then* B_S *has a natural structure of left* $\mathrm{Cart}_S(K)$*-module.*

This is a consequence of lemma (6.2), where we put $T = P$. Indeed, as $E_{P,S} \subset \mathrm{Cart}(K)$ satisfies $F_p E_{P,S} \subset E_{P,S} F_p$ for any $p \in U$, we have $E_{P,S} \cdot B_S \subset B_S$. As $I_{P,S} \cdot B_S = 0$, B_S is a left $(E_{P,S}/I_{P,S})$-module, i.e. a left $\mathrm{Cart}_S(K)$-module. Here we will allow ourselves to drop the superscripts that were used in section (6).

7.4 Definition. An S-typical group over the basic ring K is a formal group G over K, together with a $\mathrm{Cart}_S(K)$-module of S-typical curves, $\mathfrak{C}_S(G) \subset \mathfrak{C}(G)$, such that any $\gamma \in \mathfrak{C}(G)$ has a unique expansion

7.5 $$\gamma = \Sigma_{\alpha \in \underline{\underline{U}}} V_\alpha \cdot \gamma_\alpha \quad , \text{ where } \quad \gamma_\alpha \in \mathfrak{C}_S(G) \quad \text{for all } \alpha \in \underline{\underline{U}}.$$

7.6 A S-typical group G contains basic sets of S-typical curves: if $(\gamma_i)_{i \in I}$ is any basic set, we write $\gamma_i = \Sigma_\alpha V_\alpha \cdot \gamma_{\alpha,i}$ as in (7.5), then we replace each γ_i by $\gamma_{1,i} \in \mathfrak{C}_S(G)$ to obtain the basic set $(\gamma_{1,i})_{i \in I}$ of S-typical curves.

7.7 If $(\gamma_i)_{i \in I}$ is a basic set of curves in the S-typical group G over K, with $\gamma_i \in \mathfrak{C}_S(G)$ for every $i \in I$, then every $\gamma \in \mathfrak{C}(G)$ has a unique expansion as

$$\gamma = \Sigma_{\alpha \in \underline{\underline{U}}, n \in \underline{\underline{S}}, i \in I} V_{\alpha n}[x_{\alpha,n,i}] \cdot \gamma_i \quad ,$$

where $(x_{\alpha,n,i})_{i \in I} \in K^{(I)}$ for every $(\alpha,n) \in \underline{\underline{U}} \times \underline{\underline{S}}$, so that the components γ_α in (7.5) are given by

7.8 $$\gamma_\alpha = \Sigma_{n \in \underline{\underline{S}}, i \in I} V_n[x_{\alpha,n,i}] \cdot \gamma_i \in \mathfrak{C}_S(G) \ .$$

7.9 If $\gamma \in \mathfrak{C}_S(G)$, then $\gamma_1 = \gamma$ and $\gamma_\alpha = 0$ for $\alpha > 1$, and formula (7.8) says that $\mathfrak{C}_S(G)$ is a reduced $\mathrm{Cart}_S(K)$-module (with its topology of subspace of $\mathfrak{C}(G)$).

7.10 In most of the S-typical groups we shall actually study, $\mathfrak{C}_S(G)$ will be the set of all S-typical curves in G. Then any formal group homomorphism of S-typical groups, $u : G' \to G$, must map $\mathfrak{C}_S(G')$ into $\mathfrak{C}_S(G)$, i.e. $u \circ \gamma \in \mathfrak{C}_S(G)$ for any

$\gamma \in \mathfrak{C}_S(G')$. But there are some obnoxious cases (for instance if K is of characteristic $p \in U$), and that is why we had to formulate the definition of S-typical groups as in (7.4).

7.11 Definition. A morphism of S-typical groups, $u : G \to G'$, is a formal group homomorphism such that $u \circ \gamma \in \mathfrak{C}_S(G')$ for any $\gamma \in \mathfrak{C}_S(G)$.

In all but the obnoxious cases (7.10), the last condition can be dropped, so that the category of S-typical groups will be a full subcategory of that of formal groups.

7.12 Theorem. The category of S-typical groups over a basic ring K is equivalent to the category of reduced $\mathrm{Cart}_S(K)$-modules.

Proof. To any S-typical group G corresponds the $\mathrm{Cart}_S(K)$-module $\mathfrak{C}_S(G)$, which is reduced (see 7.9).

Let C be any uniform $\mathrm{Cart}_S(K)$-module and B the topological product $B = C^{\underline{\underline{U}}}$. We define the left action of $\mathrm{Cart}(K)$ on B via the ring homomorphism $\kappa_{P,S}\colon \mathrm{Cart}(K) \to \tilde{M}_{\underline{\underline{U}}}(\mathrm{Cart}_S(K))$ of theorem (6.6). More precisely, if $\kappa_{P,S}(x)_{\alpha,\beta}$ are the entries of the matrix $\kappa_{P,S}(x)$ for an $x \in \mathrm{Cart}(K)$, and $\gamma = (\gamma_\alpha)_{\alpha \in \underline{\underline{U}}} \in B$, then $x \cdot \gamma = \gamma'$, $\gamma' = (\gamma'_\alpha)_{\alpha \in \underline{\underline{U}}}$, and

7.13 $$\gamma'_\alpha = \Sigma_{\beta \in \underline{\underline{U}}} \kappa_{P,S}(x)_{\alpha,\beta} \cdot \gamma_\beta \quad , \quad \text{for any } \alpha \in \underline{\underline{U}} \ .$$

Then we can write $\gamma = \Sigma_{\alpha \in \underline{\underline{U}}} V_\alpha \cdot \gamma_\alpha$ instead of $\gamma = (\gamma_\alpha)_{\alpha \in \underline{\underline{U}}}$, by identifying any $\gamma_1 \in C$ with $(\gamma_1, 0, 0, \ldots) \in B$. On the other hand, the action of $\mathrm{Cart}(K)$ on B is the only one for which C, identified with a subgroup of B, is S-typical (with the given action of $\mathrm{Cart}_S(K)$ on C) and for which we have $\gamma = \Sigma_{\alpha \in \underline{\underline{U}}} V_\alpha \cdot \gamma_\alpha$ Theorem (6.6) is only used to insure the existence of B qua $\mathrm{Cart}(K)$-module.

Now the order function on B is given by

7.14
$$\mathrm{ord}(\Sigma_{\alpha\in\underline{\underline{U}}} V_\alpha \cdot \gamma_\alpha) = \min_{\alpha\in\underline{\underline{U}}} \alpha\ \mathrm{ord}(\gamma_\alpha) ,$$

where $\mathrm{ord}(\gamma_\alpha)$ is the order of γ_α in the uniform $\mathrm{Cart}_S(K)$-module C . It follows readily from (7.14) that <u>the</u> Cart(K)-<u>module</u> B <u>is reduced iff the</u> $\mathrm{Cart}_S(K)$-<u>module</u> C <u>is reduced</u>. In that case, theorem (III.11.25) proves the existence of a formal group G , with (up to unique isomorphism) $\mathfrak{C}(G) = B$, and we put $\mathfrak{C}_S(G) = C$ to obtain the S-typical group G .

7.15 <u>Changes of rings for</u> S-<u>typical groups</u>. Let $\varphi : K \to K'$ be a basic ring homomorphism, and G an S-typical group over K . Then the formal group $\varphi_* G$ over K' has a natural structure of S-typical group over K', by taking for $\mathfrak{C}_S(\varphi_* G)$ the closed $\mathrm{Cart}_S(K')$-module generated by all curves $\varphi_*\gamma$, $\gamma \in \mathfrak{C}_S(G)$.

The above assertions are easily checked by taking a basic set in $\mathfrak{C}_S(G)$ and applying φ_* to it (see also, later, V.6.22).

7.16 <u>The</u> S-<u>typical group</u> $\hat{W}_S^+$ is defined from the functor in groups W_S^+ , i.e. $K \mapsto$ the additive group of the ring $W_S(K)$, just as $\hat{W}^+$ was defined from W^+ (see III.1.16). <u>There is one canonical curve</u> γ_W <u>in</u> $\hat{W}_S^+$, <u>and the set of curves</u> $(F_n \cdot \gamma_W)_{n\in\underline{\underline{S}}}$ <u>is a basic set of</u> S-<u>typical curves</u>.

7.17 The $\mathrm{Cart}_S(K)$-module $\mathfrak{C}_S(\hat{W}_S^+)$ is naturally isomorphic with $\mathrm{Cart}_S(K)$, while $\mathfrak{C}(\hat{W}_S^+)$ is naturally isomorphic with a factor module of Cart(K), the kernel being the left closed ideal generated by all F_p , $p \in U$ (see 6.11).

7.18 <u>The representation theorem for</u> S-<u>typical groups</u> (see III.4.1). <u>Let</u> G <u>be an</u> S-<u>typical group over some basic ring</u> K , <u>and</u> $\gamma \in \mathfrak{C}_S(G)$. <u>Then there is exactly one</u> S-<u>typical group morphism</u> (7.11), $u_\gamma : \hat{W}_S^+ \to G$, <u>such that</u>

7.19 $$u_\gamma \circ \gamma_W = \gamma \quad .$$

This follows from theorem (7.12), as $\mathfrak{C}_S(\hat{W}_S^+) = \mathrm{Cart}_S(K)$.

7.20 The endomorphism ring of $\hat{W}_S^+$ (qua S-typical group) is $\mathrm{Cart}_S(K)$, acting on the right.

Proof: see (III.5.12).

8. The reduction theorem

8.1 The reduction theorem. Let K be a basic ring and S the set of the primes which are not invertible in K. Then any formal group G over K has a unique structure of S-typical group, $\mathfrak{C}_S(G)$ being the set of all S-typical curves $\gamma \in \mathfrak{C}(G)$.

The theorem will be obtained as a consequence of the following proposition

8.2 Proposition. Notations being as in theorem (6.6), let us assume that any prime $p \in U$ is invertible in K. Then any integer $\alpha \in \underline{\underline{U}}$ is invertible in $\mathrm{Cart}_T(K)$ (see 4.13) and the ring homomorphism $\kappa_{T,S} : \mathrm{Cart}_T(K) \to \tilde{M}_{\underline{\underline{U}}}(\mathrm{Cart}_S(K))$ is an isomorphism.

The reciprocal images $\varrho_{\alpha,\beta} \in \mathrm{Cart}_T(K)$ of the matrix units $\underline{\underline{e}}_{\alpha,\beta}$ (i.e. the elements such that $\kappa_{T,S}(\varrho_{\alpha,\beta}) = \underline{\underline{e}}_{\alpha,\beta}$) are given by the following formulas:

8.3 $$\varrho_{1,1} = \Sigma_{\alpha \in \underline{\underline{U}}}\, \alpha^{-1}\mu(\alpha) V_\alpha F_\alpha = \Pi_{p \in U}(1-p^{-1}V_p F_p) \quad ,$$

where μ denotes the Möbius function from number theory;

8.4 $$\varrho_{\alpha,\beta} = \beta^{-1} V_\alpha \varrho_{1,1} F_\beta \quad , \quad \text{for any} \quad \alpha,\beta \in \underline{\underline{U}} \quad .$$

Proof. One can prove directly from (6.10) that $\kappa_{T,S}$ is an isomorphism, but we will rather prove first (8.3) and (8.4). Indeed, we have by (6.7) and (6.9)

8.5 $$\kappa_{T,S}(V_\alpha F_\alpha) = \Sigma_{\lambda\in\underline{\underline{U}}} \alpha \underline{\underline{e}}_{\lambda\alpha,\lambda\alpha} ,$$

whence

$$\kappa_{T,S}(\Sigma_{\alpha\in\underline{\underline{U}}} \alpha^{-1}\mu(\alpha)V_\alpha F_\alpha) = \Sigma_{\alpha,\lambda\in\underline{\underline{U}}} \mu(\alpha)\underline{\underline{e}}_{\lambda\alpha\ \lambda\alpha} = \underline{\underline{e}}_{1,1} ,$$

by applying the classical property of the Möbius function, namely

8.6 $$\Sigma_{d|n}\ \mu(d) = \begin{cases} 1 & \text{if } n=1 , \\ 0 & \text{if } n > 1,\ n \in \underline{\underline{P}} . \end{cases}$$

The equality

8.7 $$\Sigma_{\alpha\in\underline{\underline{U}}} \alpha^{-1}\mu(\alpha)V_\alpha F_\alpha = \Pi_{p\in U}(1-p^{-1}V_p F_p)$$

comes from the direct definition of the Möbius function μ : if $n \in \underline{\underline{P}}$ is the product of r distinct primes, then $\mu(n) = (-1)^r$, whereas $\mu(n) = 0$ when n has a nontrivial square divisor.

So we have proved that $\kappa_{T,S}(\varrho_{1,1}) = \underline{\underline{e}}_{1,1}$ for $\varrho_{1,1} \in \mathrm{Cart}_T(K)$ as in (8.3). As (6.7) and (6.9) give

8.8 $$\kappa_{T,S}(V_\alpha)\underline{\underline{e}}_{1,1}\ \kappa_{T,S}(F_\beta) = \beta\ \underline{\underline{e}}_{\alpha,\beta} ,$$

for any $\alpha,\beta \in \underline{\underline{U}}$, we have $\kappa_{T,S}(\varrho_{\alpha,\beta}) = \underline{\underline{e}}_{\alpha\ \beta}$ for $\varrho_{\alpha,\beta} \in \mathrm{Cart}_T(K)$ defined by (8.4).

In order to prove that $\kappa_{T,S}$ is an isomorphism, it suffices to check the equality

8.9 $$\Sigma_{\alpha\in\underline{\underline{U}}}\ \varrho_{\alpha,\alpha} = 1 \quad \text{in} \quad \mathrm{Cart}_T(K) .$$

From (8.3) and (8.4) we obtain

$$\Sigma_{\alpha\in\underline{\underline{U}}}\ \varrho_{\alpha,\alpha} = \Sigma_{\alpha,\beta\in\underline{\underline{U}}}\ (\alpha\beta)^{-1}\ \mu(\beta)V_{\alpha\beta}F_{\alpha\beta} ,$$

so that (8.9) follows from the property of the Möbius function (8.6).

8.10 The reduction theorem (8.1) is just a special case of proposition (8.2), namely when $T = P$. If G is a formal group over K, and $\gamma \in \mathfrak{C}(G)$, there is exactly one S-typical curve $\gamma' \in \mathfrak{C}(G)$ such that $\mathfrak{T}\gamma' = \mathfrak{T}\gamma$, namely $\gamma' = \varepsilon_{1,1}\gamma$ or

8.11 $$\gamma' = \Sigma_{\alpha \in \underline{\underline{U}}} \alpha^{-1}\mu(\alpha)V_\alpha F_\alpha \cdot \gamma = \Sigma_{p \in U}(1-p^{-1}V_pF_p)\cdot\gamma \quad , \; U = P - S.$$

8.12 In the "<u>local case</u>", S contains exactly one prime, say p. That happens when K is a <u>ring of characteristic</u> p (i.e. $p1_K = 0$), or when K is a <u>local ring with residual field of characteristic</u> p.

9. <u>An example: between Artin-Hasse and multiplication</u>

9.1 Let S be any set of prime. We will presently define a one-dimensional group law μ_S over $\underline{\underline{Z}}$, i.e. a formal series $\mu_S(x,y) = x + y + \ldots \in \underline{\underline{Z}}[[x,y]]$. The formal group corresponding to μ_S (on the model D) will be denoted by $\mathfrak{G}_S$ and called the S-<u>typical multiplicative group</u>. The identity of D qua element of $\mathfrak{C}(\mathfrak{G}_S)$ will be denoted by γ_S.

The relations defining $\mathfrak{G}_S$ are

9.2 $$F_p \cdot \gamma_S = \gamma_S \qquad \text{if } p \in S ,$$

9.3 $$F_p \cdot \gamma_S = 0 \qquad \text{if } p \in P - S .$$

9.4 The existence and unicity of μ_S will appear as a special case of a proposition to be proved in the next chapter (see V.9.17); but it can be established by a direct argument. Namely it is apparent from (9.3) that $\mathfrak{G}_S$ is an S-typical group, so that we just have to find the corresponding reduced $\mathrm{Cart}_S(\underline{\underline{Z}})$-module. We take in $\mathrm{Cart}_S(\underline{\underline{Z}})$ the closed left ideal N_S generated by the elements $F_p - 1$, $p \in S$. Then one has to check that the factor module $\mathrm{Cart}_S(\underline{\underline{Z}})/N_S$ has the required properties, γ_S being the residual class of $1 \in \mathrm{Cart}_S(K)$.

Two special cases have already been previously denoted:

9.5 if $S = \emptyset$, then $\mathfrak{G}_{\emptyset} = \underline{\underline{G}}_a$ and $\gamma_{\emptyset} = \gamma_{\underline{\underline{a}}}$ $\mu_{\emptyset}(x,y) = x + y$ (see II.2.13);

9.6 if $S = P$, then $\mathfrak{G}_P = \underline{\underline{G}}_m$ and $\gamma_P = \gamma_{\underline{\underline{m}}}$ $\mu_P(x,y) = x + y - xy$ (see II.2.14).

9.7 Let S and T be two sets of primes, with $S \subset T$, and let us denote by $\underline{\underline{Z}}_{T-S}$ the subring of $\underline{\underline{Q}}$ generated by the elements p^{-1} , for $p \in T - S$. We are going to show that $\mathfrak{G}_S$ <u>and</u> $\mathfrak{G}_T$ <u>are isomorphic over</u> $\underline{\underline{Z}}_{T-S}$. More precisely, we shall define formal series

9.8
$$\begin{cases} ex_{T,S}(t) = t + \ldots \in \underline{\underline{Z}}_{T-S}[[t]] , \\ lo_{S,T}(t) = t + \ldots \in \underline{\underline{Z}}_{T-S}[[t]] , \end{cases}$$

which are reciprocal formal group isomorphism, $ex_{T,S}\colon \mathfrak{G}_S \to \mathfrak{G}_T$, $lo_{S,T}\colon \mathfrak{G}_T \to \mathfrak{G}_S$ (in analogy with the logarithm and the exponential). That means that

9.9
$$\begin{cases} \mu_T(ex_{T,S}(x),\ ex_{T,S}(y)) = ex_{T,S}(\mu_S(x,y)) , \\ \mu_S(lo_{S,T}(x),\ lo_{S,T}(y)) = lo_{S,T}(\mu_T(x,y)) . \end{cases}$$

Indeed $ex_{T,S} \circ \gamma_S$ has to be a curve in $\mathfrak{C}(\mathfrak{G}_T)$, over the basic ring $\underline{\underline{Z}}_{T-S}$, which satisfies the same relations as γ_S , i.e. (9.2) and (9.3). The general method would give

$$ex_{T,S} \circ \gamma_S = \Pi_{p \in U}(1-p^{-1}V_pF_p) \cdot \gamma_T = \Sigma_{\alpha \in \underline{\underline{U}}} \alpha^{-1}\mu(\alpha)V_\alpha F_\alpha \cdot \gamma_T ,$$

where $U = T - S$. But, as γ_T is invariant by the operators $F_\alpha, \alpha \in \underline{\underline{U}}$, we can as well write

9.10
$$ex_{T,S} \circ \gamma_S = \Pi_{p \in U}(1-p^{-1}V_p) \cdot \gamma_T = \Sigma_{\alpha \in \underline{\underline{U}}} \alpha^{-1}\mu(\alpha)V_\alpha \cdot \gamma_T$$

The verifications can be made directly, by using the

relations $F_p(1-p^{-1}V_p) = F_p - 1$, Now $1 - p^{-1}V_p$ is invertible in $\mathrm{Cart}(\underline{\underline{Z}}_{T-S})$, namely

$$(1-p^{-1}V_p)^{-1} = \Sigma_{n\in\underline{\underline{N}}}\, p^{-n}V_{p^n} .$$

This gives us the formula defining the reciprocal isomorphism $lo_{S,T}$, namely

9.11 $$lo_{S,T}\circ \gamma_T = \Pi_{p\in U}\,(1-p^{-1}V_p)^{-1}\circ \gamma_S = \Sigma_{\alpha\in\underline{\underline{U}}}\, \alpha^{-1}V_\alpha\cdot\gamma_S .$$

From formulas (9.10) and (9.11) we have

9.12 $$lo_{S,T}\circ ex_{T,S} = ex_{T,S}\circ lo_{S,T} = Id ,$$

moreover

9.13 $$lo_{S_1,S_2}\circ lo_{S_2,S_3} = lo_{S_1,S_3} \quad\text{and}\quad ex_{S_3,S_2}\circ ex_{S_2,S_1} = ex_{S_3,S_1},$$

when $S_1 \subset S_2 \subset S_3$.

CHAPTER V

THE STRUCTURE THEOREM AND ITS CONSEQUENCES

1. Free uniform $\mathrm{Cart}_S(K)$-modules and types

1.1 In the present chapter, S will denote some set of primes, K some basic ring, and we shall write simply E instead of $\mathrm{Cart}_S(K)$.

Our purpose is to study presentations of reduced E-modules, more precisely to study the relations between elements of a V-basis in a reduced E-module. But it will be more convenient to work with uniform E-modules (see IV.5) and to pick up later reduced modules from among uniform modules.

1.2 We shall use generators in the algebraic-topological sense: a set $(\gamma_i)_{i\in I}$ in a uniform E-module C is said to generate C iff any $\gamma \in C$ can be written as

$$\gamma = \Sigma_{i\in I}\, x_i \cdot \gamma_i ,$$

with coefficients $x_i \in E$; when I is infinite, the set $(x_i)_{i\in I}$ has to converge towards 0 .

1.3 Corresponding to any set I , there is a free uniform E-module, say L , whose elements are in one-to-one correspondence with the sets $(x_i)_{i\in I}$ in E . We may write $\Sigma_{i\in I}\, x_i \cdot \tilde{\gamma}_i$ instead of $(x_i)_{i\in I}$, defining thereby the free generators $(\tilde{\gamma}_i)_{i\in I}$ of L . The morphisms $f : L \to C$ into any uniform E-module C (see IV.5.7) are in one-to-one correspondence with the indexed sets $(\gamma_i)_{i\in I}$ in C , by the formula $f(\Sigma_{i\in I}\, x_i \cdot \tilde{\gamma}_i) = \Sigma_{i\in I}\, x_i \cdot \gamma_i$.

By (IV.2.3), every $\gamma \in L$ has a unique expansion as

1.4 $$\gamma = \Sigma_{m,n\in \underline{\underline{S}}, i\in I}\, V_m[x_{m,n,i}]F_n \cdot \tilde{\gamma}_i \quad , \quad x_{m,n,i} \in K .$$

The finiteness condition is that, for any given $m \in S$, almost all $x_{m,n,i}$ vanish.

1.5 It is clear from (1.4) that a free uniform E-module L is reduced, its natural V-basis being the set $(F_n \cdot \gamma_i)_{n\in \underline{\underline{S}}, i\in I}$.

1.6 Definition. We call "type" of a non-zero element $\gamma \in L$, and denote by $tp(\gamma)$, the pair (m_o, n_o) , where m_o is the order of γ and n_o the largest integer such that there is an $i \in I$ with $x_{m_o, n_o, i} \neq 0$ in (1.4). That is:

1.7 $tp(\gamma) = (m_o, n_o)$ iff $x_{m,n,i} = 0$ for $m < m_o$, or for $m = m_o$ and $n > n_o$; and there is $i \in I$ with $x_{m_o, n_o, i} \neq 0$. We put $tp(0) = (+\infty, 0)$.

The definition of types holds for elements of E (considered as a free E-module with generator 1_E).

1.8 We order the set of possible types, and more generally the set $\underline{\underline{P}} \times \underline{\underline{N}}$, by putting

$(m,n) < (m',n')$ iff $m < m'$ or $(m = m'$ and $n > n')$.

1.9 Lemma. Let $(\gamma_n)_{n \in \underline{\underline{N}}}$ be a sequence in L with strictly increasing types: $tp(\gamma_{n+1}) > tp(\gamma_n)$, $n \in \underline{\underline{N}}$. Then (γ_n) converges towards 0 in L .

Indeed, the sequence of orders, $ord(\gamma_n)$, is non-decreasing, and cannot remain bounded, because the bound would be reached and then we would have an infinite decreasing sequence in the well-ordered set $\underline{\underline{N}}$.

1.10 For any $(m,n) \in \underline{\underline{S}} \times \underline{\underline{N}}$, the relation $tp(\gamma) \geq (m,n)$, or $tp(\gamma) > (m,n)$, defines a closed additive subgroup of L .

That follows from (IV.2.6) and (IV.2.13).

1.11 If $tp(\gamma) = (m,n)$, then $tp(V_\mu[c] \cdot \gamma) \geq (\mu m,n)$ for any $\mu \in \underline{\underline{S}}$, $c \in K$.

That follows from (IV.2.9) and (IV.2.10).

1.12 Lemma. For any $n \in \underline{\underline{S}}$, $i \in I$, let $\gamma_{n,i} \in L$ be such that $tp(\gamma_{n,i} - F_n \cdot \tilde{\gamma}_i) > (1,n)$.

Then the set $(\gamma_{n,i})_{n \in \underline{\underline{S}}, i \in I}$ is a V-basis of L . Moreover, if the element γ of (1.4) is written as

1.13 $$\gamma = \Sigma_{m,n \in \underline{\underline{S}}, i \in I} V_m[y_{m,n,i}] \cdot \gamma_{n,i} ,$$

then (1.7) holds with $x_{m,n,i}$ replaced by $y_{m,n,i}$; i.e. the type of γ is computable from the $y_{m,n,i}$ exactly as from the $x_{m,n,i}$ (see 1.6).

1.14 Proof. Let us write $\mathfrak{L}$ instead of $gr_1 L$, and denote by $\mathfrak{x}$ the natural map $L \to \mathfrak{L}$. Then, by (IV.5.17), $(\gamma_{n,i})_{n \in \underline{\underline{S}}, i \in I}$ is a V-basis of L iff $(\mathfrak{x}\gamma_{n,i})_{n \in \underline{\underline{S}}, i \in I}$ is a basis of the free K-module $\mathfrak{L}$, with basis $(\mathfrak{x}F_n \cdot \tilde{\gamma}_i)_{n \in \underline{\underline{S}}, i \in I}$.

For any $\nu \in \underline{\underline{S}}$, let us denote by $(\mathfrak{L})_\nu$ the sub-K-module of $\mathfrak{L}$ generated by $\mathfrak{F}_n \cdot \tilde{\gamma}_i$, $n < \nu$, $i \in I$. Then we have an increasing sequence of submodules

1.15 $$0 = (\mathfrak{L})_1 \subset \dots \subset (\mathfrak{L})_n \subset (\mathfrak{L})_{n+1} \subset \dots ,$$

whose union is $\mathfrak{L}$. The hypothesis is expressed by the relations

1.16 $$\mathfrak{\gamma}_{n,i} - \mathfrak{F}_n \cdot \tilde{\gamma}_i \in (\mathfrak{L})_n .$$

They imply, by induction on $\nu \in \underline{\underline{S}}$, that $(\mathfrak{L})_\nu$ is freely generated both by the $\mathfrak{F}_n \cdot \tilde{\gamma}_i$ or by the $\mathfrak{\gamma}_{n,i}$ for $n < \nu$, $i \in I$. By (1.16) and (1.10), we have $tp(\gamma_{n,i}) = (1,n)$.

The first assertion is therefore proved, and the element γ of (1.4) has a unique expansion as (1.13). Let (m_1,n_1) be defined by the $y_{m,n,i}$ as (m_o,n_o) was defined in (1.7) by the $x_{m,n,i}$. We have

1.17 $$\gamma - \Sigma_{i\in I} V_{m_1}[y_{m_1,n_1,i}]F_{n_1} \cdot \tilde{\gamma}_i = \gamma' + \gamma'' ,$$

where

$$\gamma' = \Sigma_{m,n\in\underline{\underline{S}},(m,n)>(m_1,n_1),i\in I} V_m[y_{m,n,i}] \cdot \gamma_{n,i} ,$$

$$\gamma'' = \Sigma_{i\in I} V_{m_1}[y_{m_1,n_1,i}] \cdot (\gamma_{n_1,i} - F_{n_1} \cdot \tilde{\gamma}_i) .$$

Now properties (1.10), (1.11), joint to the hypothesis, imply that $tp(\gamma') > (m_1,n_1)$, $tp(\gamma'') > (m_1,n_1)$, and (1.17) implies $tp(\gamma) = (m_1,n_1)$.

2. Slopes and types

2.1 With the notations of section (1) and of theorem (IV.3.18), let $L_Q \subset L$ denote the set of $\gamma = \Sigma_{i\in I} x_i \cdot \tilde{\gamma}_i$ with $x_i \in E_Q$ for any $i \in I$. In other words, if

$$\gamma = \Sigma_{m,n,i} V_m[x_{m,n,i}]F_n \cdot \tilde{\gamma}_i , \quad x_{m,n,i} \in K ,$$

then $\gamma \in L_\varrho$ means that $x_{m,n,i} = 0$ unless $m/n = \varrho$.

The following assertions are easy consequences of theorem (IV.3.18).

<u>Any</u> $\gamma \in L$ <u>has a unique expansion as a converging sum</u>

2.2 $$\gamma = \Sigma_{\varrho \in \mathrm{Sl}\, S}\, \gamma_\varrho \quad , \quad \gamma_\varrho \in L_\varrho \quad .$$

2.3 <u>For any</u> $\varrho, \sigma \in \mathrm{Sl}\ S$, L_σ <u>is a closed additive subgroup of</u> L , <u>and</u>

2.4 $$E_\varrho \cdot L_\sigma \subset L_{\varrho\sigma} \quad .$$

We are now introducing two relations, namely

2.5 $$\gamma \geqslant_{sl} \alpha \quad , \text{ and}$$

2.6 $$\gamma >_{sl} \alpha \quad .$$

(read: as for slopes, γ...), where $\gamma \in L$ and α is a real positive number. <u>If</u> γ <u>is as in</u> (2.2), <u>then</u> (2.5) <u>means that</u> $\gamma_\varrho = 0$ <u>for</u> $\varrho < \alpha$, <u>and</u> (2.6) <u>means that</u> $\gamma_\varrho = 0$ <u>for</u> $\varrho \leqslant \alpha$. This definition holds for $x \in E$ (instead of $\gamma \in L$) by considering E as a special case of L .

2.7 <u>Remark</u>. In the local case (see IV.8.12), slopes are only powers p^h , $h \in \underline{\underline{Z}}$, of some prime p ; but in the other cases, Sl S is a dense subgroup of the multiplicative group $\underline{\underline{Q}}_+^\times$, and there is a significant difference between (2.5) and (2.6).

2.8 <u>For a given</u> α , <u>either of relations</u> (2.5), (2.6) <u>defines a closed subgroup of</u> L .

2.9 <u>The relations</u> $x \geqslant_{sl} \alpha$ <u>and</u> $\gamma \geqslant_{sl} \beta$, <u>where</u> $x \in E$, $\gamma \in L$, $\alpha, \beta > 0$, <u>imply</u> $x \cdot \gamma \geqslant_{sl} \alpha\beta$, <u>and</u> $x \cdot \gamma >_{sl} \alpha\beta$ <u>if only</u> $x >_{sl} \alpha$ <u>or</u> $\gamma >_{sl} \beta$.

The above assertions follow from (2.2), (2.3).

2.10 For an element $\gamma \in L$, $\gamma \neq 0$, there is a "best possible" relation of the kind (2.5) or (2.6). Indeed, for

$$\gamma = \Sigma_{m,n\in\underline{\underline{S}},i\in I} V_m[x_{m,n,i}]F_n\cdot\tilde{\gamma}_i \text{ , put}$$

$\alpha = \text{glb}(m/n)$ for such m,n as exists i with $x_{m,n,i} \neq 0$. Then the best relation is $\gamma \geq_{sl} \alpha$ if the lower bound α is reached (i.e. if α is a minimum), or $\gamma >_{sl} \alpha$ if it is not.

2.11 Lemma. Let elements $\gamma_{n,i} \in L$ be given, such that

2.12 $$(\gamma_{n,i} - F_n\cdot\tilde{\gamma}_i) >_{sl} n^{-1} \text{ , for any } n \in \underline{\underline{S}} \text{ , } i \in I \text{ .}$$ Then the hypothesis of lemma (1.12) is fulfilled. Moreover, if

2.13 $$\gamma = \Sigma_{m,n\in\underline{\underline{S}},i\in I} V_m[y_{m,n,i}]\cdot\gamma_{n,i} \quad , \quad y_{m,n,i} \in K \text{ ,}$$

than the best relation of the kind (2.5) or (2.6) for γ is defined by the support of the set $(y_{m,n,i})_{m,n\in S,i\in I}$ exactly as it was defined by the support of the set $(x_{m,n,i})_{m,n\in\underline{\underline{S}},i\in I}$ in (2.10).

Proof. For any $\gamma \in L$ and $n \in \underline{\underline{S}}$, the relation $\gamma >_{sl} n^{-1}$ implies $tp(\gamma) > (1,n)$ or equivalently $\mathfrak{T}\gamma \in (\mathfrak{T}L)_n$ with the notation of (1.15). Therefore we may apply the conclusion of lemma (1.12) and write any $\gamma \in L$ in the form (2.13). Put

$\alpha = \text{glb}(m/n)$ for such m,n as exists i with $y_{m,n,i} \neq 0$.

First let us assume that α is a minimum. Then we have $\gamma \geq_{sl} \alpha$ by (2.8), (2.9), and we have to show that $\gamma \not>_{sl} \alpha$ (i.e. that $\gamma >_{sl} \alpha$ does not hold). Let m_o be the smallest m for which there are n_o, i such that $m_o/n_o = \alpha$, $y_{m_o,n_o,i} \neq 0$. Put $\gamma = \gamma' + \gamma''$ with

2.14 $$\gamma' = \Sigma_{m,n\in\underline{\underline{S}},m<m_o,i\in I} V_m[y_{m,n,i}]\cdot\gamma_{n,i} \quad ,$$

2.15 $$\gamma'' = \Sigma_{m,n\in\underline{\underline{S}},m\geq m_o,i\in I} V_m[y_{m,n,i}]\cdot\gamma_{n,i} \quad .$$

Then we have $\gamma' >_{sl} \alpha$ and, by lemma (1.12), $tp(\gamma'') = (m_o, n_o)$, so that $\gamma'' \not>_{sl} \alpha$, and $\gamma \not>_{sl} \alpha$.

Now let us assume that α is not a minimum. Then we have $\gamma >_{sl} \alpha$, and we have to show that it is the best relation, i.e. that $\gamma \not\geq_{sl} \beta$ for $\beta > \alpha$. Here we define m_o as the smallest m for which there are n_o, i, with $m_o/n_o < \beta$ and $y_{m_o,n_o,i} \neq 0$. We put $\gamma = \gamma' + \gamma''$, with γ', γ'' as in (2.14), (2.15). We have $\gamma' \geq_{sl} \beta$, and, by lemma (1.12), $tp(\gamma'') \leq (m_o, n_o)$, so that $\gamma'' \not\geq_{sl} \beta$ and $\gamma \not\geq_{sl} \beta$.

2.16 Lemma. Let $x \in E$ and $\gamma \in L$ be such that $tp(x) = (m,n)$ and $\gamma >_{sl} \nu^{-1}$ for some $\nu \in \underline{\underline{S}}$. Then $tp(x \cdot \gamma) > (m, n\nu)$.

Proof. By (1.10), it suffices to prove that

$tp(V_m[a]F_n \cdot \gamma) > (m, n\nu)$, for any $m,n \in \underline{\underline{S}}$, $a \in K$. But we have

$$V_m[a]F_n \geq_{sl} m/n \quad \text{and} \quad \gamma >_{sl} \nu^{-1},$$

therefore, by (2.9),

2.17 $$V_m[a]F_n \cdot \gamma >_{sl} m(n\nu)^{-1},$$

and, as $ord(V_m[a]F_n \cdot \gamma) \geq m$, (2.17) implies $tp(V_m[a]F_n \cdot \gamma) > (m, n\nu)$.

3. The structure theorem

3.1 The structure theorem. Let L be the reduced E-module freely generated by the set $(\tilde{\gamma}_i)_{i \in I}$ (see 1.3). Let elements $u_{p,i} \in L$ be given, such that

3.2 $$u_{p,i} >_{sl} p^{-1}, \quad \text{for any } p \in S,\ i \in I.$$

<u>Put</u>

3.3 $$\varepsilon_{p,i} = F_p \cdot \tilde{\gamma}_i - u_{p,i} \quad ,$$

<u>denote by</u> N <u>the closed submodule of</u> L <u>with generators</u> $\varepsilon_{p,i}$ $(p \in S,\ i \in I)$, <u>by</u> C <u>the factor module</u> L/N <u>and by</u> $\gamma_i \in C$ <u>the natural image of</u> $\tilde{\gamma}_i \in L$ $(i \in I)$.

<u>Define as follows the elements</u> $\gamma_{n,i} \in L$, <u>for</u> $n \in \underline{\underline{S}}$, $i \in I$:

3.4 $$\gamma_{1,i} = \tilde{\gamma}_i \quad , \text{ for any } i \in I$$

3.5 $$\gamma_{n,i} = F_{n/q} \cdot \varepsilon_{q,i} \quad , \text{ for } i \in I,\ n \in \underline{\underline{S}}\ ,\ n > 1\ ,$$

q <u>denoting the smallest prime divisor of</u> n.

<u>Then</u> $(\gamma_{n,i})_{n \in \underline{\underline{S}}, i \in I}$ <u>is a</u> V-<u>basis of</u> L, <u>and any</u> $\gamma \in L$ <u>has a unique expansion</u>

3.6 $$\gamma = \Sigma_{m,n \in \underline{\underline{S}}, i \in I} V_m[y_{m,n,i}] \cdot \gamma_{n,i} \quad .$$

<u>Let</u> $N_o \subset N$ <u>be the subset defined by the conditions</u> $y_{m,1,i} = 0$ <u>for any</u> $m \in \underline{\underline{S}}$, $i \in I$, <u>in formula</u> (3.6). <u>Then the following four assertions are equivalent, and are always verified in the local case</u> (see IV.8.12).

3.7 A <u>The uniform module</u> C <u>is reduced, with the</u> V-<u>basis</u> $(\gamma_i)_{i \in I}$;

3.8 B $N_o = N$;

3.9 C <u>for any</u> $\gamma \in N,\ \gamma \neq 0$, $tp(\gamma) < (ord(\gamma), 1)$;

3.10 D <u>for any</u> $p, q \in S,\ i \in I$, <u>there are elements</u> $v_{p,q,i,r,j}$ (<u>with</u> $r \in S,\ j \in I$), <u>such that</u>

3.11 $$v_{p,q,i,r,j} >_{sl} r(pq)^{-1} \quad , \text{ for any } p,q,r \in S\ ,\ i,j \in I\ ,$$

<u>and</u>

3.12 $$F_p\, \varepsilon_{q,i} - F_q \cdot \varepsilon_{p,i} = \Sigma_{r \in S, j \in I}\, v_{p,q,i,r,j}\ \varepsilon_{r,j} \quad .$$

Proof. We have

$$F_n\cdot\tilde{\gamma}_i - \gamma_{n,i} = \begin{cases} 0 & \text{if } n = 1 , \\ F_{n/q}u_{q,i} & \text{if } n > 1 . \end{cases}$$

From (3.2) and (2.9), we obtain

3.13 $$\gamma_{n,i} - F_n\cdot\tilde{\gamma}_i >_{sl} n^{-1} , \text{ for any } n \in \underline{\underline{S}}, i \in I .$$

Therefore we may apply lemmas (1.12) and (2.11); the set $(\gamma_{n,i})_{n\in\underline{\underline{S}},i\in I}$ is a V-basis of L.

Let the subset $N_1 \subset L$ be defined by the conditions $y_{m,n,i} = 0$ for $n > 1$ and any $m \in \underline{\underline{S}}$, $i \in I$, in formula (3.6), so that any $\gamma \in L$ can be written as $\gamma = \gamma_o + \gamma_1$, $\gamma_o \in N_o$, $\gamma_1 \in N_1$.

The factor module $C = L/N$ (with its order topology) admits the V-basis $(\gamma_i)_{i\in I}$ iff the restriction to N_1 of the natural map $L \to C$ is bijective, and we know already that it is surjective. It is injective iff $N_o = N$; indeed, if $N_o = N$, then N_1 is a full set of representatives of L modulo N, and, if $N_o \neq N$, then $N \cap N_1 \neq 0$.

It follows from lemma (1.12) that, for any $\gamma \in N_o$, $\gamma \neq 0$, we have $tp(\gamma) < (ord(\gamma),1)$, while for any $\gamma \in N_1$, $\gamma \neq 0$, we have $tp(\gamma) = (ord(\gamma),1)$. We have proved the equivalence of A, B and C.

<u>In the local case</u>, there is only one prime in S, and we obtain $N_o = N$ by writing any element of N in the form $\Sigma_{i\in I}\, x_i\, \varepsilon_{p,i}$, $x_i \in E$, and expanding the x_i.

Let us put

$$\delta_{p,q,i} = F_p\cdot\varepsilon_{q,i} - F_q\cdot\varepsilon_{p,i} , \quad \text{for } p,q \in S,\ i \in I.$$

Then

$$\delta_{p,q,i} = F_q\cdot u_{p,i} - F_p\cdot u_{q,i} ,$$

so that, by (3.2) and (2.8), (2.9),

3.14 $$\delta_{p,q,i} >_{sl} (pq)^{-1} .$$

Clearly $\delta_{p,q,i} \in N$, so that, if $N = N_o$, we have expansions

3.15 $$\delta_{p,q,i} = \Sigma_{m,n\in\underline{\underline{S}},n>1,j\in I} V_m[z_{p,q,i,m,n,j}]\gamma_{n,j} .$$

It follows from (3.14) and lemma (2.11) that $z_{p,q,i,m,n,j} = 0$ in (3.15) when $m/n \leq (pq)^{-1}$, so that we obtain (3.12) by replacing, in (3.15), $\gamma_{n,j}$ by its value (3.5), and B(3.8) implies D(3.10).

3.16 Lemma. Let $\gamma = \Sigma_{p\in S,i\in I}\, x_{p,i}\cdot\varepsilon_{p,i}$ $(x_{p,i} \in E)$ be an element of N . Then

3.17 $$tp(\gamma) \geq \min_{p\in S,i\in I} tp(x_{p,i}F_p) .$$

Moreover, if D(3.10) holds, and if

3.18 $$tp(\gamma) > \min_{p\in S,i\in I} tp(x_{p,i}F_p)$$

then there is another set of coefficients $y_{p,i} \in E$, such that

3.19 $$\Sigma_{p\in S,i\in I}\, y_{p,i}\cdot \varepsilon_{p,i} = \Sigma_{p\in S,i\in I}\, x_{p,i}\cdot \varepsilon_{p,i} \text{ , and}$$

3.20 $$\min_{p\in S,i\in I} tp(y_{p,i}F_p) > \min_{p\in S,i\in I} tp(x_{p,i}F_p) .$$

This lemma will show that assertion D(3.10) implies C(3.9), and therefore will complete the proof of the structure theorem. For, by lemma (1.9), the process of lemma (3.16), i.e. the replacement of the $x_{p,i}$ by the $y_{p,i}$, cannot be repeated infinitely if $\gamma \neq 0$, so that there must be equality in some relation of the kind (3.17), whence clearly $tp(\gamma) < (ord(\gamma),1)$.

Proof of lemma (3.16). By (3.2) and lemma (2.16), we have

3.21 $$tp(x\cdot u_{p,i}) > tp(x\, F_p) \text{ , for any } x \in E \text{ , } x \neq 0,\ p \in S,$$
$i \in I$,

so that $tp(x.\varepsilon_{p,i}) = tp(x\ F_p)$, whence (3.17).

Now let us assume that (3.18) holds, and put

3.22 $$(m,n) = \min_{p\in S, i\in I} tp(x_{p,i}F_p) .$$

We have $n \in \underline{\underline{S}}$, $n > 1$. Our first step is to get rid of the $x_{p,i}$ for which the minimum in (3.22) is not reached, and to replace the other ones by "monomials". More precisely, we define the elements $x'_{p,i} \in E$ <u>such that</u> $tp(x'_{p,i}F_p) > (m,n)$, for any $p \in S$, $i \in I$, by putting

3.23 $$\begin{cases} x_{p,i} = x'_{p,i} & \text{if } p \nmid n , \\ x_{p,i} = x'_{p,i} + V_m[a_{p,i}]F_{n/p} & \text{if } p|n , \quad a_{p,i} \in K . \end{cases}$$

Then (3.18) means that

3.24 $$tp(\Sigma_{p|n,i\in I} V_m[a_{p,i}]F_n\cdot\tilde{\gamma}_i) > (m,n) ,$$

which is equivalent, by (IV.2.13), to

3.25 $$\Sigma_{p|n}\ a_{p,i} = 0 .$$

Let q be the smallest prime divisor of n . If $p|n$, $p \neq q$, then $pq|n$, and we may write

3.26 $$\begin{aligned} F_{n/p}\cdot\varepsilon_{p,i} &= F_{n/pq}\cdot(F_q\cdot\varepsilon_{p,i}) \\ &= F_{n/pq}(F_q\cdot\varepsilon_{p,i} - F_p\cdot\varepsilon_{q,i}) + F_{n/q}\cdot\varepsilon_{q,i} \end{aligned}$$

Let us put

3.27 $$\begin{cases} x''_{p,i} = 0 & \text{if } p \neq q \\ x''_{q,i} = \Sigma_{p|n} V_m[a_{p,i}]F_{n/q} . \end{cases}$$

By (3.25), we have $tp(x''_{p,i}\ F_p) > (m,n)$ for all $p \in S$, $i \in I$. By (3.23), (3.26), (3.27), we have

3.28 $$\Sigma_{p\in S,i\in I} (x_{p,i}-x'_{p,i}-x''_{p,i})\cdot\ \varepsilon_{p,i} =$$

$$\Sigma_{p|n,p\neq q,i\in I} V_m[a_{p,i}]F_{n/pq}(F_q\cdot\varepsilon_{p,i}-F_p\cdot\varepsilon_{q,i}) .$$

By (3.12), we have

$$F_q \cdot \varepsilon_{p,i} - F_p \cdot \varepsilon_{q,i} = \Sigma_{r\in S, j\in I} V_{q,p,i,r,j} \cdot \varepsilon_{r,j} ,$$

so that, by permuting the indices p,r and i,j ,

3.29 $$\Sigma_{p\in S, i\in I}(x_{p,i} - x'_{p,i} - x''_{p,i}) \cdot \varepsilon_{p,i} = \Sigma_{p\in S, i\in I} x'''_{p,i} \cdot \varepsilon_{p,i} , \text{ where}$$

3.30 $$x'''_{p,i} = \Sigma_{r|n, r\neq q, j\in I} V_m[a_{r,j}]F_{n/qr} v_{q,r,j,p,i} .$$

As, by (3.11), $v_{q,r,j,p,i} >_{sl} p(qr)^{-1}$, so that $v_{q,r,j,p,i} F_p >_{sl} (qr)^{-1}$ and, by (3.30) and lemma (2.16),

3.31 $$tp(x'''_{p,i} F_p) > (m,n) .$$

Relations (3.19) and (3.20) are therefore verified, by putting

$$y_{p,i} = x'_{p,i} + x''_{p,i} + x'''_{p,i} , \quad \text{for any } p \in S, \ i \in I .$$

4. A second proof of the existence theorem

4.1 By the "existence theorem", we mean the statement that to any reduced $\mathrm{Cart}_S(K)$-module C there corresponds an S-typical group G , such that $\mathfrak{C}_S(G)$ is isomorphic to C .

We proved it before in the case where $S = P$ (see III.11.25); then we treated the case of any set of primes S , by reduction to the former case (see IV.7.12).

We are going to give another proof of this crucial result, using only the first part of the structure theorem, i.e. we need only the equivalence of assertions A and B in theorem (3.1). For this, we have to explain some general notions concerning embedded subgroups and factor groups.

4.2 Consider, over a basic ring K , a formal variety G split as a direct product : $G = V \times W$. For brevity's sake, we shall speak of points in G , meaning points in $G(A)$ for some $A \in \underline{\underline{\mathrm{nil}}}(K)$.

A point in G is a pair (x,y), with $x \in V$, $y \in W$. The morphisms $x \mapsto (x,0)$, $V \to G$ and $y \mapsto (0,y)$, $W \to G$ are the "natural injections", and sometimes it is convenient to identify V,W with their images $V \times 0$ and $0 \times W$.

4.3 Let a formal group structure be defined on G by the group morphism $\mu : G \times G \to G$. Instead of $\mu((x_1,y_1), (x_2,y_2))$, the "sum" for the operation μ of two points (x_1,y_1) and (x_2,y_2) in G will be written as

4.4 $$(x_1,y_1) +_\mu (x_2,y_2) .$$

Let us assume that the sum $(0,y_1) +_\mu (0,y_2)$ has always 0 as its first component, for any two points y_1,y_2 in W. This is expressed by writing

4.5 $$(0,y_1) +_\mu (0,y_2) = (0,\mu'(y_1,y_2)) ,$$

where μ' is a morphism, $\mu' : W \times W \to W$. One verifies immediately that μ' is a group morphism (see II.2.3), so that (4.5) may be rewritten as

4.6 $$(0,y_1) +_\mu (0,y_2) = (0,y_1 +_{\mu'} y_2) .$$

It is not a serious restriction to assume that

4.7 $(x,y) = (x,0) +_\mu (0,y)$ for any point (x,y) in G.

Indeed, the morphism $V \times W \to G$, $(x,y) \mapsto (x,0) +_\mu (0,y)$ is an isomorphism (see I.8.1), so that the property (4.7) may be assumed, without changing anything to $V \times 0$ nor to $0 \times W$. We applied this argument before, when introducing _curvilinear coordinates_ (see 11.7).

The morphism $V \times V \to V \times W$, $(x_1,x_2) \mapsto (x_1,0) +_\mu (x_2,0)$ is defined by its two component morphisms, $\mu'' : V \times V \to V$ and $f : V \times V \to W$, such that

4.8 $$(x_1,0) +_\mu (x_2,0) = (\mu''(x_1,x_2)\ ,\ f(x_1,x_2))\ .$$

Now, assuming (4.7) to hold, we have

4.9 $$(x_1,y_1) +_\mu (x_2,y_2) = (\mu''(x_1,x_2),\ y_1 +_{\mu'} y_2 +_{\mu'} f(x_1,x_2))\ .$$

Indeed:

$$\begin{aligned}(x_1,y_1)+_\mu (x_2,y_2) &= (x_1,0)+_\mu(0,y_1)+_\mu(x_2,0)+_\mu(0,y_2)\\ &= (x_1,0)+_\mu(x_2,0)+_\mu(0,y_1)+_\mu(0,y_2)\\ &= (\mu''(x_1,x_2),0)+_\mu(0,f(x_1,x_2))+_\mu(0,y_1)+_\mu(0,y_2)\\ &= (\mu''(x_1,x_2),0)+_\mu(0,f(x_1,x_2)+_{\mu'}y_1+_{\mu'}y_2)\\ &= (\mu''(x_1,x_2),y_1+_{\mu'}y_2+_{\mu'}f(x_1,x_2))\ .\end{aligned}$$

Then it is easy to check that μ'' is a group morphism on V, so that the group morphism μ is defined by two group morphisms, μ', μ'', and a morphism f, according to the formula.

4.10 $$(x_1,y_1) +_\mu (x_2,y_2) = (x_1+_{\mu''}x_2,\ y_1+_{\mu'}y_2+ f(x_1,x_2))\ .$$

By checking the associativity of μ, we obtain the condition

$$f(x_1,x_2) +_{\mu'} f(x_1+_{\mu''}x_2,x_3) = f(x_2,x_3) +_{\mu'} f(x_1,x_2+_{\mu''}x_3)\ ,$$

which may be interpreted by saying that f must be 2-cocycle:

4.11 $$\delta f(x_1,x_2,x_3) = f(x_2,x_3) -_{\mu'} f(x_1+_{\mu''}x_2,x_3)+_{\mu'}f(x_1,x_2+_{\mu''}x_3)-_{\mu'} f(x_1,x_2) = 0.$$

4.12 Conversely, two groups morphisms μ', μ'' and a 2-cocycle f define a group morphism μ by formula (4.10), because (4.11) expresses the associativity of μ, and implies $f(x_1,0) = f(0,x_2) = 0$, so that $\mathfrak{T}f = 0$ and the condition on $\mathfrak{T}\mu$ is verified (see II.2.4).

4.13 <u>Definition</u>. <u>An embedded subgroup of an</u> S-<u>typical group</u> G is an S-typical group G', such that:

4.14 for any $A \in \underline{\underline{nil}}(K)$, $G'(A)$ is a subgroup of $G(A)$;

4.15 there is a formal variety V and an isomorphism $u : V \times G' \to G$, such that $u_A(0,y) = y$ for any $y \in G'(A)$, $A \in \underline{\underline{nil}}(K)$;

4.16 $\mathfrak{C}_S(G') = \mathfrak{C}(G') \cap \mathfrak{C}_S(G)$ (as pointed out in (IV.7.10), this last condition is generally superfluous).

4.17 Proposition. Notation being as in (4.13), there is a unique S-typical group, denoted by $G'' = G/G'$, such that, for any $A \in \underline{\underline{nil}}(K)$, $G''(A) = G(A)/G'(A)$ (the morphism $G \to G''$ being a morphism of S-typical groups).

This results from our previous considerations; G'' is isomorphic to the formal variety V together with the group morphism μ''.

4.18 Proof of the existence theorem. The S-typical group $\hat{W}_S^+$ (see IV.7.16) may be defined directly (see III.2.15), and the representation theorem (see IV.7.18), $\mathfrak{C}_S(G) \simeq \mathrm{Hom}_S(\hat{W}_S^+, G)$, may be deduced from theorem (III.4.1). Indeed, in formula (III.4.4), the terms with $n \notin \underline{\underline{S}}$ may be omitted if they vanish.

4.19 Free uniform (or reduced) $\mathrm{Cart}_S(K)$-modules correspond to direct sums of isomorphic copies of $\hat{W}_S^+$.

4.20 Let us go back to theorem (3.1) and denote by W the S-typical group such that (up to unique isomorphism), $\mathfrak{C}_S(W) = L$. Let J be the set of pairs (n,i), where $n \in \underline{\underline{S}}$, $n > 1$, $i \in I$, and write $\tilde{\gamma}_j = \gamma_{n,i}$ for $j = (n,i)$. So we have a basic set of curves in $\mathfrak{C}_S(W)$: the $\tilde{\gamma}_i$, $i \in I$, and the $\tilde{\gamma}_j$, $j \in J$. They define a curvilinear group law, say μ, on the model

$D^{(I\cup J)} = D^{(I)} \times D^{(J)}$. The only thing that has to be verified in order to apply proposition (4.17) and obtain the group law μ'' on $D^{(I)}$ (see 4.10), is that $0 \times D^{(J)}$ is an embedded subgroup of $D^{(I)} \times D^{(J)}$. For this we consider the S-typical group W_1, such that $\mathbb{C}_S(W_1)$ is freely generated by curves $\tilde{\gamma}_{p,i}$, $p \in S$, $i \in I$, and the S-typical group morphism $r : W_1 \to W$, defined by $r \circ \tilde{\gamma}_{p,i} = \varepsilon_{p,i}$ for any $p \in S$, $i \in I$. Then condition (3.8) can be interpreted by saying that, for any $A \in \underline{\underline{nil}}(K)$, $0 \times A^{(J)} \subset A^{(I\cup J)}$ is the image of the group homomorphism r_A . The S-typical group G such that $\mathbb{C}_S(G) = C$ is defined by the curvilinear group law μ'' corresponding to the basic set $(\gamma_i)_{i\in I}$.

5. Presentations, structural constants, morphisms

5.1 Let G be an S-typical group over K and $(\gamma_i)_{i\in I}$ a basic set in $\mathbb{C}_S(G)$. Let us choose, for any $p \in S$, $i \in I$, elements $x_{p,i,j} \in \mathrm{Cart}_S(K)$, $j \in I$, such that

5.2 $$F_p \cdot \gamma_i = \Sigma_{j\in I}\, x_{p,i,j} \cdot \gamma_j \quad , \quad i \in I ,$$

5.3 $$x_{p,i,j} >_{sl} p^{-1} \qquad \text{(see 2.6)} .$$

Then we have a presentation of G , i.e. a definition by generators (the curves γ_i , or if one prefers, the elements $\gamma_i(t)$ for $t \in A \in \underline{\underline{nil}}(K)$) and relations (5.3). Indeed, it follows from the structure theorem (3.1) that any relation between the γ_i is a consequence of relations (5.2). The S-typical group G is defined, up to unique isomorphism, by its presentation.

Equivalently, the morphisms of G into an S-typical group G' are in one-to-one correspondence with the indexed sets

$(\gamma'_i)_{i\in I}$ in $\mathbb{C}_S(G')$, such that

5.4 $$F_p\cdot\gamma'_i = \Sigma_{j\in J}\ x_{p,i,j}\cdot\gamma'_j\quad,\ \text{for any}\ i\in I,\ p\in S\ .$$

The corresponding morphism $f : G \to G'$ is defined by

5.5 $$f\circ\gamma_i = \gamma'_i\quad,\ \text{for any}\ i\in I\ .$$

5.6 For instance, the morphisms $\underline{\underline{G}}_a \to G$ correspond to curves $\gamma\in\mathbb{C}_S(G)$ such that $F_p\cdot\gamma = 0$ for any $p\in S$, and the morphisms $\Phi_S \to G$ (see IV.9) to curves $\gamma\in\mathbb{C}_S(G)$ such that $F_p\cdot\gamma = \gamma$ for any $p\in S$.

The S-typical group $\hat{W}_S^+$ has a natural basic set $(\gamma_i)_{i\in I}$, with relations $F_p\cdot\gamma_i = \gamma_{pi}$. One may say that the canonical curve $\gamma_W = \gamma_1$ is "free", and that accounts for the universal property of $\hat{W}_S^+$ (see IV.7.18).

Expanding the coefficients $x_{p,i,j}\in \mathrm{Cart}_S(K)$ in (5.2) as in (IV.2.3), we have

5.7 $$x_{p,i,j} = \Sigma_{m,n\in\underline{\underline{S}}}\ V_m[c_{m,n,p,i,j}]F_n\quad,\quad c_{m,n,p,i,j}\in K$$

and condition (5.3) means that

5.8 $$c_{m,n,p,i,j} = 0\quad\text{unless}\quad m/n > p^{-1}\ .$$

It is always possible to take $c_{m,n,p,i,j} = 0$ if $n > 1$ in (5.7). Given the basic set $(\gamma_i)_{i\in I}$ in $\mathbb{C}_S(G)$, the elements $c_{m,p,i,j}\in K$ are defined by the relations

5.9 $$F_p\cdot\gamma_i = \Sigma_{m\in\underline{\underline{S}},j\in I}\ V_m[c_{m,p,i,j}]\cdot\gamma_i\quad,\quad p\in S,\ i\in I\ ,$$

and we call them the <u>structural constants of</u> G, relative to the γ_i (by analogy with the structural constants $c_{i,j,k}$ of an algebra).

5.10 The structural constants $c_{m,p,i,j}$ do depend on the basic set $(\gamma_i)_{i\in I}$, and cannot be chosen arbitrarily, except in the

<u>local case</u>.

When one has to change the basic set $(\gamma_i)_{i\in I}$, it is easier to allow coefficients $x_{p,i,j}$ of the form $\Sigma_{m\in\underline{\underline{S}}} V_m \operatorname{op}(\xi_m)$, where $\xi_m \in W_S(K)$: see (IV.4.1). It amounts to requiring that, in (5.7), $c_{m,n,p,i,j} = 0$ unless $m/n \in \underline{\underline{P}}$, which obviously implies (5.8).

5.11 The elements $\Sigma_{m\in\underline{\underline{S}}} V_m \operatorname{op}(\xi_m)$ in $\operatorname{Cart}_S(K)$ are the ones that are <u>sums of elements of integral slope</u>. Therefore, by corollary (IV.3.23), they form a <u>subring</u> of $\operatorname{Cart}_S(K)$, <u>which we shall denote by</u> $\operatorname{Cart}_S(K)_+$.

Addition and multiplication in $\operatorname{Cart}_S(K)_+$ are defined by the following formulas, where $(\xi_m)_{m\in\underline{\underline{S}}}$ and $(\eta_m)_{m\in\underline{\underline{S}}}$ are indexed sets in $W_S(K)$, subject to no restriction.

5.12 $$\Sigma_{m\in\underline{\underline{S}}} V_m \operatorname{op}(\xi_m) + \Sigma_{m\in\underline{\underline{S}}} V_m \operatorname{op}(\eta_m) = \Sigma_{m\in\underline{\underline{S}}} V_m \operatorname{op}(\xi_m+\eta_m) ;$$

5.13 $$\begin{cases} (\Sigma_{m\in\underline{\underline{S}}} V_m \operatorname{op}(\xi_m))(\Sigma_{m\in\underline{\underline{S}}} V_m \operatorname{op}(\eta_m)) = \Sigma_{m\in\underline{\underline{S}}} V_m \operatorname{op}(\zeta_m) \text{, where} \\ \zeta_m = \Sigma_{n|m} (F_n \cdot \xi_{m/n})\, \eta_n \qquad \text{(see IV.4.19)} . \end{cases}$$

5.14 In the <u>local case</u>, i.e. when $S = \{p\}$ for some prime p, condition (5.3) means that the coefficients $x_{p,i,j}$ are to be chosen in $\operatorname{Cart}_{\{p\}}(K)_+$. The subscript p may be dropped, so that we write simply V, F instead of V_p, F_p. Then $V_p h = V^h$, $F_p h = F^h$ for any $h \in \underline{\underline{N}}$ (see IV.2.9). The ring $W_{\{p\}}(K)$ is the ring of "classical" Witt vectors; elements of $\operatorname{Cart}_{\{p\}}(K)_+$ are "<u>twisted formal series in</u> V" (see 5.13).

Let us repeat that the conditions of the existence theorem (3.1) are always verified in the local case, and state it as a theorem.

5.15 <u>Theorem. Let</u> I <u>be any set, and</u> $\xi_{i,h,j}$, $h \in \underline{\underline{N}}$, $i,j \in I$,

<u>be elements of</u> $\text{Cart}_{\{p\}}(K)$, <u>subject to no condition if</u> I <u>is finite, and to the condition that</u> $(\xi_{i,h,j})_{j\in I}$ <u>converges towards</u> 0 <u>for any</u> $h \in \underline{\underline{N}}$ <u>if</u> I <u>is infinite</u>.

<u>Then the presentation</u>

5.16 $$F\cdot\gamma_i = \Sigma_{h\in\underline{\underline{N}},j\in I} V^h \text{op}(\xi_{i,h,j})\cdot\gamma_j \quad , \quad i \in I ,$$

<u>defines a</u> {p}-<u>typical group</u> G <u>over</u> K, <u>with the basic set</u> $(\gamma_i)_{i\in I}$ <u>of</u> {p}-<u>typical curves. Moreover, the presentation</u> (5.16) <u>can be replaced by an equivalent presentation</u>

5.17 $$F\cdot\gamma_i = \Sigma_{h\in\underline{\underline{N}},j\in I} V^h[c_{i,h,j}]\cdot\gamma_j \quad , \quad i \in I ,$$

<u>where the</u> $c_{i,h,j}$ <u>are elements of</u> K, <u>defined by the basic set</u> $(\gamma_i)_{i\in I}$ <u>in</u> G, <u>and called the structural constants of</u> G <u>relative to</u> $(\gamma_i)_{i\in I}$.

6. <u>Tensor products</u>

6.1 Let us put $E = \text{Cart}_S(K)$, for some set of primes S and some basic ring K .

As a consequence of the representation theorem for S-typical groups, the ring E may be identified with the endomorphism ring $\text{End}(\hat{W}_S^+)$ acting on the <u>right</u>. It means that, for any $A \in \underline{\underline{\text{nil}}}(K)$, the additive group $\hat{W}_S^+(A)$ has a natural structure of <u>right</u> E-module.

6.2 Let us denote by e_x the endomorphism of $\hat{W}_S^+$ corresponding to $x \in E$. Then e_x is <u>defined</u> by

6.3 $$e_x \circ \gamma_W = x\cdot\gamma_W \in \mathfrak{C}_S(\hat{W}_S^+) .$$

Therefore

6.4 $$e_x \circ (F_n\cdot\gamma_W) = F_n\cdot(e_x \circ \gamma_W) = F_n\, x\cdot\gamma_W \quad \text{for any } n \in \underline{\underline{S}} .$$

If $A \in \underline{\underline{nil}}(K)$, a point $a \in \hat{W}^+_S(A)$ is written as a <u>finite</u> sum: $a = \Sigma_{n \in \underline{\underline{S}}} (F_n \cdot \gamma_w)(a_n)$ or as an indexed set, $a = (a_n)_{n \in \underline{\underline{S}}} \in A^{(\underline{\underline{S}})}$. Therefore we have, by (6.4)

6.5 $$a \cdot x = \Sigma_{n \in \underline{\underline{S}}} (F_n x \cdot \gamma_w)(a_n) \in \hat{W}^+_S(A) \ .$$

6.6 <u>For any</u> $A \in \underline{\underline{nil}}(K)$, $\hat{W}^+_S(A)$, <u>together with its discrete topology, is a topological right E-module</u>.

It means that, for any $a \in \hat{W}^+_S(A)$, we have $a \cdot x = 0$ for $x \in E$, provided $\mathrm{ord}(x)$ is large enough.

Let $a_n \in A$ and $\alpha_n \in \underline{\underline{N}}$ be such that $a_n^{\alpha} = 0$ for $\alpha > \alpha_n$, $\alpha \in \underline{\underline{P}}$. Then $(y \cdot \gamma)(a_n) = 0$ for any curve $\gamma \in \mathfrak{C}_S(G)$ and any $y \in E$ such that $\mathrm{ord}(y) > \alpha_n$. By (IV.2.18), we have $\mathrm{ord}(F_n \cdot x) \geqslant n^{-1}\mathrm{ord}(x)$, whence $a \cdot x = 0$ in (6.5) provided that $n^{-1}\mathrm{ord}(x) > \alpha_n$ for every $n \in \underline{\underline{S}}$, or rather for the n such that $a_n \neq 0$ (i.e. $\alpha_n > 0$), which are in finite number.

6.7 <u>Lemma</u>. <u>For any</u> S-<u>typical group</u> G <u>and any</u> $A \in \underline{\underline{nil}}(K)$, <u>there is a map</u>

6.8 $$\varphi_{G,A}: \hat{W}^+_S(A) \otimes_E \mathfrak{C}_S(G) \to G(A) \ ,$$

<u>defined by</u>

6.9 $$\varphi_{G,A}(a \otimes \gamma) = u_{\gamma,A}(a) \ , \ a \in \hat{W}^+_S(A) \ , \quad \gamma \in \mathfrak{C}_S(G)$$

<u>where</u> $u_\gamma : \hat{W}^+_S \to G$ <u>is the morphism defined by</u>

6.10 $$u_\gamma \circ \gamma_W = \gamma \ .$$

<u>The map</u> $\varphi_{G,A}$ <u>is additive, and is functorial both in</u> G <u>and in</u> A.

Proof. As $u_{\gamma,A}(a)$ is additive both in $a \in \hat{W}^+_S(A)$ and in $\gamma \in \mathfrak{C}_S(G)$, it remains only to check that

6.11 $$u_{\gamma,A}(a \cdot x) = u_{x \cdot \gamma,A}(a) \quad , \quad \text{for any } x \in E \ .$$

By (6.10), we have

$$x\cdot\gamma = x\cdot(u_\gamma \circ \gamma_W) = u_\gamma \circ (x\cdot\gamma_W) = u_\gamma \circ e_x \circ \gamma_W$$

(by 6.3), so that $u_{x\cdot\gamma} = u_\gamma \circ e_x$, which is formula (6.11).

As a special case of (6.9), we have

6.12 $$\varphi_{G,A}(\gamma_W(t)\otimes\gamma) = \gamma(t) , \quad \text{for any } t \in A,\ \gamma \in \mathfrak{C}_S(G) ,$$

whence it follows that $\varphi_{G,A}$ is surjective.

When $G = \hat{W}_S^+$, then $\mathfrak{C}_S(G) = E\cdot\gamma_W$, and any element of $\hat{W}_S^+(A)\otimes_E E\cdot\gamma_W$ is written uniquely as $a\otimes\gamma_W$, $a \in \hat{W}_S^+(A)$. Putting $a = \Sigma_{n\in\underline{\underline{S}}}(F_n\cdot\gamma_W)(a_n)$, a <u>finite</u> sum, we obtain

6.13 $$\begin{aligned}(a\otimes\gamma_W) &= \Sigma_{n\in\underline{\underline{S}}}(F_n\cdot\gamma_W)(a_n)\otimes\gamma_W \\ &= \Sigma_{n\in\underline{\underline{S}}}\gamma_W(a_n)\otimes F_n\cdot\gamma_W ,\end{aligned}$$

therefore, by (6.12),

6.14 $$\varphi_{\hat{W}_S^+}(a\otimes\gamma) = \Sigma_{n\in\underline{\underline{S}}}(F_n\cdot\gamma_W)(a_n) = a .$$

6.15 So, $\varphi_{G,A}$ <u>is a bijection when</u> $G = \hat{W}_S^+$, and also when G is the direct sum of a finite set of copies of $\hat{W}_S^+$.

Let L be a free uniform E-module with infinite generating set $(\tilde{\gamma}_i)_{i\in I}$ (see 1.3). Then any $\gamma \in L$ has a unique expansion as $\Sigma_{i\in I}x_i\cdot\tilde{\gamma}_i$, with coefficients $x_i \in E$ converging towards 0.

On the other hand, L corresponds to an infinite direct sum W of copies of $\hat{W}_S^+$ (see 4.19). So, the map $\varphi_{W,A}$ would be a bijection if we had, for $a \in \hat{W}_S^+(A)$,

6.16 $$a\otimes(\Sigma_{i\in I}x_i\cdot\tilde{\gamma}_i) = \Sigma_{i\in I}a\cdot x_i\otimes\tilde{\gamma}_i ,$$

which would make sense, because $a\cdot x_i = 0$ for almost all $i \in I$, by (6.6).

Now (6.16) <u>holds if</u> S <u>is finite</u>, say $S = \{p_1,\dots,p_r\}$, because, if $n_1, \dots, n_r \in \underline{\underline{N}}$ and $n = p_1^{n_1} \dots p_r^{n_r}$, then any $x \in E$ with $\operatorname{ord}(x) \geq n$ can be written in the form

$$x = V_{p_1}^{n_1} x_1 + \dots + V_{p_r}^{n_r} x_r \text{ , where } x_1, \dots, x_r \in E \text{ .}$$

But, when both I and S are infinite, there is no apparent reason why (6.16) should hold and, as it must hold here, we put the following definition.

6.17 Definition. Let M be a discrete topological right E-module and C a uniform E-module (with its order topology). Then the reduced tensor product of M and C over E, denoted by $M \underline{\otimes}_E C$, is a factor group of the additive group $M \otimes_E C$; elements of the kernel are sums of elements $m \otimes \gamma$ for which there is a decomposition of $\gamma \in C$ as a converging sum, $\gamma = \Sigma_{j \in J} \gamma_j$, such that $m \otimes \gamma_j = 0$ in $M \otimes_E C$ for any $j \in J$.

6.18 Theorem. Let G be an S-typical group over K and $A \in \underline{\underline{\text{nil}}}(K)$. Then there is an additive isomorphism

6.19 $$\psi_{G,A} \colon \hat{W}_S^+(A) \underline{\otimes}_E \mathfrak{C}_S(G) \to G(A) \text{ ,}$$

defined by

6.20 $$\psi_{G,A}(\gamma_w(t) \underline{\otimes} \gamma) = \gamma(t) \text{ , for any } t \in A \text{ , } \gamma \in \mathfrak{C}_S(G) \text{ ,}$$

which is functorial in G and in A. If G is finite dimensional, or if S is finite, then $\hat{W}_S^+(A) \underline{\otimes} \mathfrak{C}_S(G) = \hat{W}_S^+(A) \otimes \mathfrak{C}_S(G)$.

Proof. The map $\psi_{G,A}$ factors the map $\varphi_{G,A}$ of (6.8). If G is any direct sum of copies of $\hat{W}_S^+$, then $\psi_{G,A}$ is a bijection: that is why we put the definition (6.17).

Now we argue as in (4.20), by introducing an exact sequence

$$\mathfrak{C}_S(W_1) \to \mathfrak{C}_S(W) \to \mathfrak{C}_S(G) \to 0 \text{ ,}$$

where W_1, W are direct sums of copies of $\hat{W}_S^+$, the corresponding sequence

$$W_1(A) \to W(A) \to G(A) \to 0$$

beeing exact for any $A \in \underline{\underline{\text{nil}}}(K)$. We have a diagram

6.21
$$\begin{array}{ccccccc} W_S^+(A)\underline{\otimes}\, \mathbb{C}_S(W_1) & \longrightarrow & W_S^+(A)\underline{\otimes}\, \mathbb{C}_S(W) & \longrightarrow & W_S^+(A)\underline{\otimes}\, \mathbb{C}_S(G) & \to & 0 \\ \psi_{W_1,A}\downarrow & & \psi_{W,A}\downarrow & & \psi_{G,A}\downarrow & & \\ W_1(A) & \longrightarrow & W(A) & \longrightarrow & G(A) & \longrightarrow & 0 \end{array}$$

The upper row in (6.21) comes from factoring the exact sequence of (ordinary) tensor products. If $\gamma = \Sigma_{j\in J}\, \gamma_j$ is a convergent decomposition in $\mathbb{C}_S(G)$, then we may lift each γ_j into $\tilde{\gamma}_j \in \mathbb{C}_S(W)$, so that the sum $\tilde{\gamma} = \Sigma_{j\in J}\, \tilde{\gamma}_j$ converges, for instance by taking $\tilde{\gamma}_j$ satisfying $\operatorname{ord}(\tilde{\gamma}_j) = \operatorname{ord}(\gamma_j)$. By this remark one proves readily that the upper row is exact, and as $\psi_{W,A}$ and $\psi_{W_1,A}$ are isomorphisms, so is $\psi_{G,A}$.

6.22 Proposition. Let

$$F_p\cdot\gamma_i = \Sigma_{j\in I}\, x_{p,i,j}\cdot\gamma_j \quad , \quad i \in I \, , \; x_{p,i,j} \in \operatorname{Cart}_S(K)$$

be a presentation of an S-typical group G, as in (5.1), and let $\varphi : K \to K'$ be a basic ring homomorphism. Then

6.23
$$F_p\cdot(\varphi_*\gamma_i) = \Sigma_{j\in I}\,(\varphi_* x_{p,i,j})\cdot(\varphi_*\gamma_j) \quad , \quad i \in I$$

is a presentation of the S-typical group $\varphi_* G$ over K' (see IV. 7.15).

Proof. It follows from (IV.1.6) and (IV.2.5) that the relations (6.23) are true in $\mathbb{C}_S(\varphi_* G)$, and we have a presentation of $\varphi_* G$, because the relations (5.3) hold for the elements $\varphi_* x_{p,i,j}$.

6.24 Corollary. The natural map $\operatorname{Cart}_S(K')\otimes_E \mathbb{C}_S(G) \to \mathbb{C}_S(\varphi_* G)$ is bijective when G is finite dimensional, or if K' is finitely generated qua K-module (for instance if φ is surjective).

7. Definition and general properties of reduced derivatives

7.1 Later, we will try to explain where the reduced derivative does come from (see section 11). Here we give a definition that would rather be a theorem in a wider frame.

7.2 Definition. Let G be an S-typical group over a basic ring K, $\mathfrak{T}G$ its tangent space and $\mathfrak{T} : \mathfrak{C}_S(G) \to \mathfrak{T}G$ the natural map. For any curve $\gamma \in \mathfrak{C}_S(G)$, we denote by $\breve{D}\gamma$ and we call the reduced derivative of γ the formal series with coefficients from $\mathfrak{T}G$,

7.3 $$\breve{D}\gamma(t) = \Sigma_{n \in \underline{\underline{S}}} t^{n-1} \mathfrak{T}(F_n \cdot \gamma) .$$

Note that the "constant term" $\mathfrak{T}\gamma$ is not generally 0.

7.4 Proposition. Let $u : G \to G'$ be a morphism of S-typical groups over K, and $\varphi : K \to K'$ a basic ring homomorphism. If, for some $\gamma \in \mathfrak{C}_S(G)$,

7.5 $$\breve{D}\gamma(t) = \Sigma_{n \in \underline{\underline{S}}} t^{n-1} a_n \quad , \quad a_n \in \mathfrak{T}G ,$$

then we have

7.6 $$\breve{D}(u \circ \gamma)(t) = \Sigma_{n \in \underline{\underline{S}}} t^{n-1} (\mathfrak{T}u) \cdot a_n \quad , \quad (\mathfrak{T}u) \cdot a_n \in \mathfrak{T}G'$$

7.7 $$\breve{D}(\varphi_* \gamma)(t) = \Sigma_{n \in \underline{\underline{S}}} t^{n-1} (\varphi_* a_n) \quad , \quad \varphi_* a_n \in \mathfrak{T}(\varphi_* G) .$$

Proof. Formula (7.6) comes from (I.6.4) and (III.3.25); formula (7.7) comes from (I.11.10) and (III.3.26).

7.8 Definitions. Let M be a K-module. We denote by $M_S[[t]]$ the additive group of formal series, $f(t) = \Sigma_{n \in \underline{\underline{S}}} t^{n-1} a_n$,

$a_n \in M$, with its usual topology (that of $M^{\underline{\underline{S}}}$, with discrete M). We let the ring $\mathrm{Cart}_S(K) = E$ act continuously on the left of $M_S[[t]]$ via the ring homomorphism κ' of (IV.3.11). More precisely, if $f(t)$ is as above, and $x \in E$, than $(x\cdot f)(t) = \Sigma_{n\in\underline{\underline{S}}} t^{n-1} a'_n$, with

7.9 $$a'_n = \Sigma_{m\in\underline{\underline{S}}} \kappa'(x)_{n,m} a_m \quad , \quad n \in \underline{\underline{S}} \qquad \text{(see IV.3.16)} ,$$

or equivalently, if $x = \Sigma_{m,n\in\underline{\underline{S}}} V_m[x_{m,n}]F_n$,

7.10 $$a'_n = \Sigma_{d|n,m\in\underline{\underline{S}}} dx^{n/d}_{d,m} a_{mn/d} \quad , \quad n \in \underline{\underline{S}} \qquad \text{(see IV.3.12)} .$$

7.11 Proposition. For any S-typical group G over K, the map $\check{D} : \mathfrak{C}_S(G) \to \mathfrak{T}G_S[[t]]$ is continuous and $\mathrm{Cart}_S(K)$-linear.

Proof. For any $\gamma \in \mathfrak{C}_S(G)$, $\mathfrak{T}\gamma = 0$ means that $\mathrm{ord}(\gamma) > 1$, so that the continuity of $\check{D}$ results from the relation $\mathrm{ord}(F_n\cdot\gamma) \geq n^{-1}\,\mathrm{ord}(\gamma)$ (see IV.2.18). In order to prove that $\check{D}(x\cdot\gamma) = x\cdot\check{D}\gamma$, it suffices to take $x \in \mathrm{Cart}_S(K)$ of the form F_m, V_m , $[c]$ and to apply formulas (IV.3.10). If $\check{D}\gamma(t) = \Sigma_{n\in\underline{\underline{S}}} t^{n-1} a_n$, we have to check that

7.12 $$\begin{cases} \check{D}(F_m\cdot\gamma)(t) = \Sigma_{n\in\underline{\underline{S}}} t^{n-1} a_{nm} , \\ \check{D}(V_m\cdot\gamma)(t) = \Sigma_{n\in\underline{\underline{S}}} t^{mn-1} ma_n , \\ \check{D}([c]\cdot\gamma)(t) = \Sigma_{n\in\underline{\underline{S}}} t^{n-1} c^n a_n , \end{cases}$$

which comes from the definition (7.3) and axioms (IV.2.9) to (IV.2.12).

7.13 Proposition. Let G be an additive group, $G = L^+$ (see II. 2.26). Then any curve $\gamma \in \mathfrak{C}_S(G)$ is written $\gamma(t) = \Sigma_{n\in\underline{\underline{S}}} t^n a_n$, and $\check{D}\gamma$ is the ordinary derivative: $\check{D}\gamma(t) = \Sigma_{n\in\underline{\underline{S}}} t^{n-1} na_n$ (see III.3.16).

In any S-typical group G, the choice of a basic set $(\gamma_i)_{i\in I}$ in $\mathfrak{C}_S(G)$ defines a basis of $\mathfrak{T}G$ over K, namely $(\mathfrak{T}\gamma_i)_{i\in I}$. The reduced derivatives $\check{D}\gamma_i$ are given by formulas

7.14 $$\check{D}\gamma_i(t) = \Sigma_{n\in\underline{\underline{S}}}\, t^{n-1} a_{i,n}\ ,\ \text{where}\ \ i\in I,\ a_{i,n}\in\mathfrak{T}G\ ,$$

and $a_{i,1} = \mathfrak{T}\gamma_i$.

Therefore every $a_{i,n}$ must be a (finite) linear combination, with coefficients from K, of the $a_{i,1}$, $i\in I$.

7.15 Proposition. Let

7.16 $$F_p\cdot\gamma_i = \Sigma_{j\in I}\, x_{p,i,j}\cdot\gamma_j\quad,\quad i\in I\ ,\quad x_{p,i,j}\in \mathrm{Cart}_S(K)\ ,$$

be a presentation of an S-typical group G, as in (5.1), and

7.17 $$x_{p,i,j} = \Sigma_{m,n\in\underline{\underline{S}}}\, V_m[c_{m,n,p,i,j}]F_n\ ,\ c_{m,n,p,i,j}\in K\ .$$

Then the coefficients $a_{i,n}\in\mathfrak{T}G$ (see 7.14) verify the relations

7.18 $$a_{i,pn} = \Sigma_{d|n,m\in\underline{\underline{S}},j\in I}\, dc^{n/d}_{d,m,p,i,j}\, a_{j,mn/d}\ ,\ i\in I,\ p\in S,\ n\in\underline{\underline{S}}\ .$$

The conditions "$c_{m,n,p,i,j} = 0$ unless $m/n > p^{-1}$" (see 5.8) imply that the $a_{i,n}\in\mathfrak{T}G$ are computable from relations (7.18) as linear combinations of the $a_{i,1}$, with coefficients in the polynomial ring $K[c_{m,n,p,i,j}]_{m,n\in\underline{\underline{S}},p\in S,i,j\in I}$.

Proof. We apply $\check{D}$ to both sides of (7.16), to obtain, by proposition (7.11),

7.19 $$F_p\cdot\check{D}\gamma_i = \Sigma_{j\in I}\, x_{p,i,j}\cdot\check{D}\gamma_j\quad \text{in}\quad \mathfrak{T}G_S[[t]]\ .$$

Then we expand the $x_{p,i,j}$ as in (7.17) and we apply formula (7.10) to the expansions (7.14) of $\check{D}\gamma_i$. Formula (7.18) is obtained by equating the coefficients of t^{n-1} in both sides of (7.19).

In (7.18), we can add in the sum on the right side the condition $d/m > p^{-1}$, i.e. $mn/d < np$. So, if $n \in \underline{\underline{S}}$, $n > 1$, we take any prime divisor p of n and we obtain $a_{i,n}$ as a linear combination of the $a_{j,n}$, where $n' < n$. In the local case, there is only one prime divisor of n, so there is no choice. But in the other cases, n may have several prime divisors, so that there are several formulas to compute one $a_{n,i}$: of course, they must give the same result, and this shows that not any presentation (5.1) defines a S-typical group, i.e. that the conditions of the structure theorem (see 3.1) are not superfluous.

7.20 Proposition. With the notation of (7.14), let $\gamma = \Sigma_{m\in\underline{\underline{S}},i\in I} V_m[x_{m,i}]\cdot\gamma_i$ be a curve in $\mathfrak{C}_S(G)$, $x_{m,i} \in K$. Then

7.21 $$\check{D}\gamma(t) = \Sigma_{n\in\underline{\underline{S}}} t^{n-1}(\Sigma_{d|n,i\in I}\, d\, x_{d,i}^{n/d}\, a_{i,n/d}),$$

by (7.10) and (7.11).

8. S-typical groups over S-torsion-free rings

8.1 Definitions. Let S be a set of primes. In the present section we consider a S-torsion-free basic ring K, i.e. a ring where the multiplication by any $p \in S$ is injective. Such a ring can be embedded in its ring of quotients $K[p^{-1}]_{p\in S}$, which we denote by K_S, with the inclusion map $\iota : K \to K_S$. When M denotes a K-module, M_S will denote the K_S-module $K_S \otimes_K M$.

8.2 Proposition. Let G be a S-typical group over K_S, and $\mathfrak{C}_\emptyset(G) \subset \mathfrak{C}_S(G)$ be the set of $\emptyset$-typical curves, or additive subgroups in $\mathfrak{C}_S(G)$. Then the restriction of $\mathfrak{T} : \mathfrak{C}_S(G) \to \mathfrak{T}G$ to $\mathfrak{C}_\emptyset(G)$ is bijective. There is a unique morphism of S-typical groups,

8.3 $$\log_G : G \to (\mathfrak{T}G)^+ ,$$

which induces identity on $\mathfrak{T}G$. For any $\gamma \in \mathfrak{C}_S(G)$, we have

8.4 $$(\log_G \circ \gamma)(t) = \Sigma_{n \in \underline{\underline{S}}} n^{-1} t^n \mathfrak{T}(F_n \cdot \gamma) ,$$

or in short

8.5 $$\log_G \circ \gamma = \int \breve{D}\gamma .$$

Proof. Here is a second application of proposition (IV.8.2), with T,S replaced respectively by $S,\emptyset$. For any $\gamma \in \mathfrak{C}_S(G)$

8.6 $$\gamma' = \Pi_{p \in S}(1-p^{-1}V_pF_p)\cdot\gamma = \Sigma_{n \in \underline{\underline{S}}} n^{-1}\mu(n)V_nF_n\cdot\gamma$$

is a curve in $\mathfrak{C}_S(G)$, such that $\mathfrak{T}\gamma' = \mathfrak{T}\gamma$ and $F_p\cdot\gamma' = 0$ for any $p \in S$, therefore for any $p \in P$; moreover, if $F_p\cdot\gamma = 0$ for any p, then $\gamma' = \gamma$ in (8.6). That shows that $\mathfrak{T} : \mathfrak{C}_\emptyset(G) \to \mathfrak{T}G$ is bijective.

Any curve $\gamma \in \mathfrak{C}_\emptyset(G)$ verifies

8.7 $$\gamma(t+t') = \gamma(t) + \gamma(t') , \text{ for any } t,t' \in A,\ A \in \underline{\underline{\mathrm{nil}}}(K) .$$

(see 5.6). The isomorphism $\log_G$ may be defined by taking a basic set of curves in $\mathfrak{C}_\emptyset(G)$. It satisfies $(\log_G\circ\gamma)(t) = t\,\mathfrak{T}\gamma$, for any $\gamma \in \mathfrak{C}_\emptyset(G)$. The more general formula (8.4) comes from (7.6), with $u = \log_G$ and $\mathfrak{T}u = \mathrm{Id}$.

8.8 Proposition (8.2) is a generalization of the $\underline{\underline{Q}}$-theorem (see II.3.1). It says that, if one has disposed beforehand of the primes outside S (by assuming that $F_p\cdot\gamma = 0$ for $\gamma \in \mathfrak{C}_S(G)$, $p \in P$, $p \notin S$) it suffices to assume that all primes inside S are invertible to obtain the equivalence between groups and modules. Note that, apart from the language, the notion of $\emptyset$-typical group was studied in (1.4) under the name of formal module. (In the case of a basic ring K of characteristic > 0, the formal module defined by K^+ has more structure than a formal <u>group</u> isomorphic to it).

8.9 The category of S-typical groups over an S-torsion free ring K will be studied as a <u>subcategory</u> of that of S-typical groups over K_S (or equivalently of that of free modules over K_S). Indeed we consider K as a subring of K_S, and when one uses models and group laws, the identification becomes obvious. But morphisms and isomorphisms over K_S are not always defined over K (if $K \neq K_S$): it is not a full subcategory. <u>The functor</u> $\mathfrak{T}$, <u>which is fully faithful over</u> K_S, <u>is only faithful over</u> K (i.e. morphisms are defined by their tangent maps, which are not, in general, arbitrary linear maps).

8.10 <u>The structure theorem over</u> S-<u>torsion-free rings. Let notations be as in theorem</u> (3.1) <u>and add the condition that the basic ring</u> K <u>is</u> S-<u>torsion-free. For every</u> $p \in S$, $i \in I$, <u>expand</u> $u_{p,i}$ <u>as</u>

$$u_{p,i} = \Sigma_{m,n\in \underline{\underline{S}},j\in I} V_m[c_{m,n,p,i,j}]F_n \cdot \tilde{\gamma}_j$$

<u>Then conditions</u> A,B,C,D <u>of theorem</u> (3.1) <u>are also equivalent to the following</u>.

8.11 E <u>In the free</u> K-<u>module</u> M <u>with basis</u> $(a_{i,1})_{i\in I}$, <u>the system of equations</u>

8.12 $$a_{i,pn} = \Sigma_{d|n,m\in \underline{\underline{S}},j\in I}\, d\, c^{n/d}_{d,m,p,i,j}\, a_{j,mn/d}\,,$$

where $i \in I$, $p \in S$, $n \in \underline{\underline{S}}$, <u>has a solution</u> $(a_{i,n})_{i\in I,n\in \underline{\underline{S}}}$.

Proof. By proposition (7.15) condition A (3.7) and the existence theorem imply condition E (8.11).

Conversely, assume that E holds, and define curves $\gamma_i^!$ in the formal module M_S^+ by

8.13 $$\gamma_i'(t) = \Sigma_{n\in\underline{\underline{S}}}\ t^n\ n^{-1}\ a_{i,n} \quad , \quad i \in I .$$

Remember that $\mathrm{Cart}(K)$ acts on $\mathbb{C}(M_S^+)$ via κ , not κ' (see IV.3.13). Then it is easy to check that the curves γ_i' satisfy the equations

8.14 $$F_p \cdot \gamma_i' = \Sigma_{m,n\in\underline{\underline{S}},j\in I}\ V_m[c_{m,n,p,i,j}]F_n \cdot \gamma_j' .$$

Therefore, if condition B (3.8) did not hold, there would be a non trivial relation of the form

8.15 $$\Sigma_{m\in\underline{\underline{S}},i\in I}\ V_m[x_{m,i}]\cdot\gamma_i' = 0 ,$$

where $(x_{m,i})_{i\in I} \in K_S^{(I)}$ for any $m \in \underline{\underline{S}}$, or equivalently

8.16 $$\Sigma_{m\in\underline{\underline{S}},i\in I}\ \gamma_i'(x_{m,i}t^m) = 0 .$$

But, if m_o is the smallest m for which not all $x_{m,i}$ vanish, the coefficient of t^{m_o} in the left side of (8.16) would be $\Sigma_{i\in I}\ x_{m_o,i}\ a_{i,1}$, in contradiction with our assumptions.

8.17 Let G be an S-typical group over K . Then G has a logarithm, <u>defined over</u> K_S , <u>not</u> K <u>in general</u> (see 8.3). Once $\log_G : G \to (\mathfrak{T}G)^+$ is known, the group morphism $f : G \times G \to G$ is defined by

8.18 $$\log_G(f(x,y)) = \log_G(x) + \log_G(y) .$$

<u>When is</u> f <u>defined over</u> K (<u>not only</u> K_S)?

The choice of a basic set $(\gamma_i)_{i\in I}$ in $\mathbb{C}_S(G)$ identifies G with the model $D^{(I)}$, and f with a curvilinear group law given by its components $(f_i)_{i\in I}$, $f_i : D^{(I)} \times D^{(I)} \to D^{(I)}$, defined by

$$\Sigma_{i\in I}\ (\gamma_i(x_i) + \gamma_i(y_i)) = \Sigma_{i\in I}\ \gamma_i(f_i(x,y)) \quad \text{(see 11.7)}.$$

Applying $\log_G$ to both sides, we obtain the formula

8.19 $$\Sigma_{i\in I} (\log_G \circ \gamma_i)(x_i) + (\log_G \circ \gamma_i)(y_i) = \Sigma_{i\in I} (\log_G \circ \gamma_i)(f_i(x,y)),$$

which is an equality of ordinary formal series, with coefficients from $\mathfrak{T}G$. We know from (8.4) that the curves $\log_G \circ \gamma_i$ must be of the form

$$(\log_G \circ \gamma_i)(t) = \Sigma_{n\in \underline{\underline{S}}} n^{-1} t^n a_{i,n} , \quad a_{i,n} \in \mathfrak{T}G .$$

8.20 Theorem. Let $(a_{i,n})_{i\in I, n\in \underline{\underline{S}}}$ be elements of a free K-module M , where K is an S-torsion-free ring, such that $(a_{i,1})_{i\in I}$ is a basis of M . Then the two following assertions are equivalent.

8.21 A There is an S-typical group G over K and a basic set $(\gamma_i)_{i\in I}$ in $\mathfrak{C}_S(G)$ such that, by identifying $\mathfrak{T}G$ with M $(\mathfrak{T}\gamma_i = a_{i,1})$,

8.22 $$\check{D}\gamma_i(t) = \Sigma_{n\in \underline{\underline{S}}} t^{n-1} a_{i,n} , \quad i \in I ,$$

or equivalently

8.23 $$(\log_G \circ \gamma_i)(t) = \Sigma_{n\in \underline{\underline{S}}} n^{-1} t^n a_{i,n} , \quad i \in I .$$

8.24 B The system of equations

8.25 $$a_{i,pn} = \Sigma_{d|n, j\in I} d\, c^{n/d}_{d,p,i,j} a_{j,n/d} ,$$

where $i \in I$, $p \in S$, $n \in \underline{\underline{S}}$, has a solution $(c_{n,p,i,j})$, $n \in \underline{\underline{S}}$, $p \in S$, $i,j \in S$, in the ring K.

Proof. Assume that A (9.21) holds. Then the S-typical group G has structural constants $c_{n,p,i,j} \in K$ corresponding to $(\gamma_i)_{i\in I}$ (see 5.9), verifying equations (8.25): indeed they are equations (8.12) where the second index of the c has been dropped.

8.26 Now assume that the $a_{i,n} \in M$ are given. Then equations (8.25) have always a unique solution in K_S: for given p and i, $c_{n,p,i,j}$ is computed by induction on n , using the fact that

$(a_{i,1})_{i\in I}$ is a basis of M. _If the_ $c_{n,p,i,j}$ _happen to lie in_ K (not only in K_S), _then we can apply the structure theorem_ (8.10), so that B implies A.

8.27 _Remark_. Theorem (8.20) makes sense in the local case, while theorem (8.10) becomes pointless (see 3.1).

8.28 _Corollary. Let the conditions of theorem_ (9.20) _be satisfied, and elements_ $(b_n)_{n\in\underline{\underline{S}}}$ _be given in_ $\mathfrak{T}G = M$. _Then there is a curve_ $\gamma \in \mathfrak{C}_S(G)$ _such that_

8.29 $$\breve{D}\gamma(t) = \Sigma_{n\in\underline{\underline{S}}}\, t^{n-1}\, b_n$$

iff the system of equations

8.30 $$b_n = \Sigma_{d|n,i\in I}\, d\, x_{d,i}^{n/d}\, a_{i,n/d} \qquad , \quad n \in \underline{\underline{S}}$$

has a solution $(x_{n,i})_{n\in\underline{\underline{S}},i\in I}$ _in_ K, _not only in_ K_S.

Proof. A curve $\gamma \in \mathfrak{C}_S(G)$ has a unique expansion as

8.31 $$\gamma = \Sigma_{n\in\underline{\underline{S}},i\in I}\, V_n[x_{n,i}]\cdot\gamma_i \quad , \quad x_{n,i} \in K$$

and, as

$$\breve{D}\gamma_i(t) = \Sigma_{n\in\underline{\underline{S}}}\, t^{n-1}\, a_{i,n} ,$$

equations (8.30) follow from formula (7.10). They have always a unique solution in K_S.

9. Some examples

9.1 Let I be any set. We shall prove the existence of a class of curvilinear group laws over $\underline{\underline{Z}}$ on the model $D^{(I)}$ and compute their logarithms. They correspond to presentations of the form

9.2 $$F_p\cdot\gamma_i = \begin{cases} V_p^{n_p}\cdot\gamma_j \\ \text{or} \\ 0 \end{cases} \qquad p \in P,\ i \in I,\ n_p \in \underline{\underline{N}} ,$$

where $j \in I$ depends on p and i. In order to simplify our formulas, let us adjoin to I some element $\omega \notin I$, and put $I' = I \cup \{\omega\}$. We make the convention that $\gamma_\omega = 0$, and we rewrite (9.2) as

9.3 $$F_p \cdot \gamma_i = V_p^{n_p} \cdot \gamma_{\sigma_p(i)} \quad , p \in P, \ i \in I ,$$

where $\sigma_p\colon I' \to I'$ is a map such that $\sigma_p(\omega) = \omega$.

9.4 <u>The presentation</u> (9.3) <u>defines a formal group iff</u> $\sigma_p \circ \sigma_q = \sigma_q \circ \sigma_p$, <u>for any two primes</u> p,q.

To prove this statement, we can check either condition D (3.10) or condition E (8.11) of the structure theorem. Let us put

9.5 $$\varepsilon_{p,i} = F_p \cdot \gamma_i - V_p^{n_p} \cdot \gamma_{\sigma_p(i)} \quad , \quad p \in P, \ i \in I', \text{ with } \varepsilon_{p,\omega} = 0 .$$

Then a computation valid in a free reduced $\text{Cart}(\underline{\underline{Z}})$-module shows that, for any two primes p,q and any $i \in I'$,

9.6 $$F_p \cdot \varepsilon_{q,i} - F_q \cdot \varepsilon_{p,i} = V_p^{n_p} \cdot \varepsilon_{q,\sigma_p(i)} - V_q^{n_q} \cdot \varepsilon_{p,\sigma_q(i)} + V_p^{n_p} V_q^{n_q} \cdot (\gamma_{\sigma_q \circ \sigma_p(i)} - \gamma_{\sigma_p \circ \sigma_q(i)}) .$$

Therefore, by condition D (3.10), the relation $\sigma_p \circ \sigma_q = \sigma_q \circ \sigma_p$ is sufficient for the existence of our group law, while it is necessary by (3.8), or by direct inspection.

Henceforth we assume that $\sigma_p \circ \sigma_q = \sigma_q \circ \sigma_p$, which enables us to define $\sigma_n\colon I' \to I'$ for any $n \in \underline{\underline{P}}$, with $\sigma_{mn} = \sigma_m \circ \sigma_n$, σ_p being as given for $p \in P$, and $\sigma_1 = \text{Id}$.

Then the equations (see 8.25)

9.7 $$a_{i,pn} = \Sigma_{d|n, j \in I}\, d\, c_{d,p,i,j}^{n/d}\, a_{j,n/d} \quad , \ i \in I', \ p \in P, \ n \in \underline{\underline{P}} ,$$

(where we put $a_{\omega,n} = 0$ for any n) take a simple form, because all structural constants $c_{n,p,i,j}$ are 0 or 1 . Namely

9.8 $$c_{n,p,i,j} = 1 \text{ if } n = p^{n_p} , j = \sigma_p(i) \text{ , and otherwise } 0 .$$

So equations (9.7) reduce to

9.9 $$a_{i,pn} = \begin{cases} 0 & \text{if } p^{n_p} \nmid n , \\ p^{n_p} a_{\sigma_p(i),np^{-n_p}} & \text{if } p^{n_p} | n . \end{cases}$$

Let us put

9.10 $$h_p = n_p + 1 \quad , \quad p \in P ,$$

so that we have

9.11 $$a_{i,p^{h_p}n} = p^{n_p} a_{\sigma_p(i),n} .$$

Let us write $n \in \underline{\underline{P}}$ as $n = \Pi_{p \in P}\, p^{\nu_p}$. Then we have

9.12 $$a_{i,n} = 0 \text{ unless } \nu_p \equiv 0 \text{ mod. } h_p \text{ for every } p \in P .$$

Let us write e_i instead of $a_{i,1}$, and $e_\omega = 0$. Then, if $n = \Pi_{p \in P}\, p^{\lambda_p h_p}$, we obtain by induction

9.13 $$a_{i,n} = \Pi_{p \in P}\, p^{\lambda_p n_p} e_{\sigma_n(i)} \quad , \quad \text{for any } i \in I .$$

We identify $(e_i)_{i \in I}$ with the natural basis of $\underline{\underline{Q}}^{(I)}$ (here $K = \underline{\underline{Z}}$ and $K_S = \underline{\underline{Q}}$), and we define the formal series ω_i by

9.14 $$\omega_i(t) = \Sigma_{n \in \underline{\underline{P}}}\, n^{-1} t^n a_{i,n} \quad , \quad \text{or}$$

9.15 $$\omega_i(t) = \Sigma_{\lambda \in \underline{\underline{N}}(P)}\, (\Pi\, p^{-\lambda_p}) t^n e_{\sigma_n(i)} \text{ , where } n = \Pi_{p \in P}\, p^{\lambda_p h_p} .$$

where products are taken with p ranging over P .

Then the curvilinear group law f we are looking for is defined by its components $f_i \colon D^{(I)} \times D^{(I)} \to D^{(I)}$ <u>with coefficients in</u> $\underline{\underline{Z}}$ (not only $\underline{\underline{Q}}$), by the formula

9.16 $$\Sigma_{i \in I}\, \omega_i(x_i) + \omega_i(y_i) = \Sigma_{i \in I}\, \omega_i(f_i(x,y)) \quad , \quad x = (x_i)_{i \in I} , \quad y = (y_i)_{i \in I} .$$

9.17 If we put $n_p = 0$ (or $h_p = 1$) for every $p \in P$, we obtain a class of formal groups which contains the Witt vectors over $\mathbb{Z}$ and their natural factor groups (see III.2.15), Barsotti's covectors and bivectors (see [1]), etc.

If we take I reduced to one element, we obtain a one-dimensional group law, defined by the coordinate curve γ such that

9.18 $$F_p \cdot \gamma = V_p^{n_p} \cdot \gamma \quad \text{or} \quad 0 \text{ , for } p \in P .$$

The corresponding S-typical group law (see theorem IV.8.1) satisfies equations (9.18) for $p \in S$, and $F_p \cdot \gamma = 0$ for $p \notin S$.

10. The parametrization of curvilinear group laws

10.1 We shall first discuss the case of dimension 1. The passage to arbitrary finite dimension will only involve more indices; infinite dimensions will bring no more (names of) indices, but a certain awkwardness in the statements.

10.2 Let us consider, over some basic ring K, a (one dimensional) group law f, i.e. a formal series

$$f(x,y) = \Sigma_{i,j \in \mathbb{N}} b_{i,j} x^i y^j \in K[[x,y]]$$

subject to certain conditions. Some of these are immediately translated into properties of the coefficients $b_{i,j}$. Namely the relation $f(x,0) = f(0,x) = x$ is translated by writing

$$f(x,y) = x + y + \Sigma_{i,j \in \mathbb{P}} b_{i,j} x^i y^j .$$

The commutativity, $f(x,y) = f(y,x)$, means that $b_{i,j} = b_{j,i}$ for any i,j, so that it suffices to consider the coefficients $b_{i,j}$ where $i \leq j$. This would lead to a formula such as

10.3 $f(x,y) = x + y + \Sigma_{1\leq i<\infty} b_{i,i}\, x^i y^i + \Sigma_{1\leq i<j<\infty} b_{i,j} (x^i y^j + x^j y^i)$.

The most important relation, associativity, i.e

10.4 $$f(f(x,y),z) = f(x,f(y,z)) ,$$

could be written down a little more explicitely, by not very much so.

However, there are some well defined polynomials with integral coefficients, in the letters $b_{i,j}$ $(1\leq i\leq j)$ denoted collectively as b , say $R_{i,j,k}$ and $R'_{i,j,k}$, such that

$$f(f(x,y),z) = \Sigma_{i,j,k} R_{i,j,k} (b)\, x^i y^j z^k ,$$

$$f(x,f(y,z)) = \Sigma_{i,j,k} R'_{i,j,k} (b)\, x^i y^j z^k .$$

Then condition (10.4) means that

10.5 $$R_{i,j,k} (b) = R'_{i,j,k} (b) , \quad \text{for any } i,j,k \in \underline{\underline{P}} .$$

Clearly, <u>there is a universal group law F over an universal ring</u> $\mathfrak{R}$: this latter has generators $B_{i,j}$, <u>subject to relations</u> (10.5) <u>and no more</u> (than their consequences): any group law over any ring K corresponds to just one ring homomorphism $\mathfrak{R} \to K$ (specializing $B_{i,j}$ in $b_{i,j}$) .

For any $q \in \underline{\underline{P}}$, the relations (10.5) where $i + j + k \leq q$ involve only the coefficients $b_{i,j}$ with $i + j \leq q$ (see I.2.4). If it is so, we say that $\Sigma_{i+j\leq q} b_{i,j} x^i y^j$ defines a q-<u>bud</u> (see II.4.1). The same argument as before proves that there is (for any q) <u>a universal</u> q-<u>bud , over an universal ring</u> $\mathfrak{R}_q$ (generated by some $B^{(q)}_{i,j}$ subject to (10.5), with $i+j+k\leq q$) , and we have ring homomorphisms

10.8 $$\mathfrak{R}_1 = \underline{\underline{Z}} \to \mathfrak{R}_2 \to \dots \to \mathfrak{R}_{q-1} \to \mathfrak{R}_q \to \dots \to \mathfrak{R} .$$

By the extension theorem (I.4.6), <u>all homomorphism in</u> (10.6)

are injective, so that we may write

10.9 $\mathfrak{R}_1 = \underline{\underline{Z}} \subset \mathfrak{R}_2 \subset \ldots \subset \mathfrak{R}_{q-1} \subset \mathfrak{R}_q \subset \ldots \subset \mathfrak{R} = \cup_q \mathfrak{R}_q$.

By lemma (II.8.1), $\mathfrak{R}_q$ is a polynomial ring in one letter (say X_q) over $\mathfrak{R}_{q-1}$; it means that the universal q-bud (over $\mathfrak{R}_q$) may be obtained by taking any q-bud over $\mathfrak{R}_{q-1}$ extending the universal (q-1)-bud, and adding $\pm X_q C_q(x,y)$.

Therefore, by induction,

$$\mathfrak{R}_q = \underline{\underline{Z}}[X_2,\ldots,X_q]$$

is a polynomial ring; q-buds over K are in one-to-one correspondence with ring homomorphisms $\mathfrak{R}_q \to K$, i.e. with specializations $X_i \mapsto x_i$ $(2 \leq i \leq q)$; moreover $x_2,\ldots,x_q$ generate the same subring of K as the coefficients $b_{i,j}$ where $i + j \leq q$.

So far, the "parameters" X_q are only defined up to transformations of the form

10.10 $$X'_q = \pm X_q + P(X_1,\ldots,X_{q-1}) \quad , \quad P : \text{a polynomial.}$$

We could fix the $\pm$ sign in (10.10), but much more can be achieved by considering the structural constants $C_{n,p}$ ($n \in \underline{\underline{P}}$, $p \in P$) of the universal group law F over the universal ring $\mathfrak{R}$. They are defined (see 5.9) as the elements $C_{n,p} \in \mathfrak{R}$ such that

10.11 $$F_p \cdot \gamma = \Sigma_{n \in \underline{\underline{P}}} V_n[C_{n,p}] \cdot \gamma ,$$

where γ is the identity curve and the sum in the right side is computed relatively to the group law F . There are universal formulas for computing the structural constants $C_{n,p}$ from the coefficients $B_{i,j}$ of F , which hold for any group law, by specialization.

The general relation $\mathrm{ord}(F_n \cdot \gamma) \geq n^{-1}\mathrm{ord}(\gamma)$ (see III.3.27) imply that the q-jet of $F_n \cdot \gamma$ depends only on the qn-jet of γ. This means, as will become more apparent below, that

10.12 $$C_{n,p} \in \mathfrak{R}_{np} \quad , \quad \text{for any } n \in \underline{\underline{P}} \ , \ p \in P \ .$$

Together with the structural constants $C_{n,p}$, we have the coefficients $A_n \in \mathfrak{R}$ $(n \in \underline{\underline{P}}\ ,\ A_o = 1)$ of the reduced derivative of the identity curve ,

10.13 $$\check{D}\gamma(t) = \Sigma_{n \in \underline{\underline{P}}}\ t^{n-1} A_n \ ,$$

with the relations (see 7.18)

10.14 $$A_{np} = n\ C_{n,p} + \Sigma_{d|n, d<n}\ d\ C_{d,p}^{n/d}\ A_{n/d} \ .$$

As $\mathfrak{R}$ is torsion free, we may write the logarithm of the universal group law F over $\mathfrak{R}$:

10.15 $$\Omega(t) = \Sigma_{n \in \underline{\underline{P}}}\ t^n n^{-1} A_n \ ,$$

10.16 $$\Omega(F(x,y)) = \Omega(x) + \Omega(y) \ , \text{ or } \ F(x,y) = \Omega^{-1}(\Omega(x) + \Omega(y)) \ .$$

By (10.16), we see that it is equivalent to give the q-bud defined by F (i.e. its coefficients $B_{i,j}$, generating $\mathfrak{R}_q$) or to give the q-jet $J_q\Omega$, i.e. its coefficients $A_2,\dots,A_q$. Even if we did not know the relations (10.12), but only the obvious relations $C_{n,p} \in \mathfrak{R}$, we would prove by (10.14) that $A_n \in \mathfrak{R}$ for any n , and by (10.16) that $A_q \in \underline{\underline{Q}} \otimes \mathfrak{R}_q$, which imply $A_q \in \mathfrak{R}_q$ for the additive group $\mathfrak{R}/\mathfrak{R}_q$ is torsion free. By the same argument and again (10.14), we would prove that $C_{n,p} \in \mathfrak{R}_{np}$.

Now let us consider equations (10.14) for a given $q = np$, assuming that we have already computed the (q-1)-bud defined by the universal group law F . There are two cases to distinguish.

10.17 First case: q is a prime power, say $q = p^{h+1}$, $h \in \underline{\underline{N}}$, $p \in P$.

Then there is only one equation to consider, namely

10.18 $$A_{p^{h+1}} = p^h C_{p^h,p} + \Sigma_{0\leq k<h}\, p^k C^{p^{h-k}}_{p^k,p} A_{p^{h-k}} .$$

By (10.18) and induction, we see that $A_{p^{h+1}}$ is a polynomial in the $C_{p^k,p}$ where $k \leq h$, more precisely that $A_{p^{h+1}} - p^h C_{p^h,p}$ depends only on the $C_{p^k,p}$ with $k < h$.

By the homomorphism obstruction lemma (III.5.2), by formulas (III.5.13, III.5.17) and the relation $p|\eta_{p^{h+1}}$ (i.e. only $(x+y)^{p^{h+1}} \equiv x^{p^{h+1}} + y^{p^{h+1}}$ mod. p) we see that <u>any choice</u> of $C_{p^h,p}$ in an extension of $\mathfrak{R}_{p^{h+1}-1}$ would define a p^{h+1}-bud, because $C_{p^h,p}$ influences the p^{h+1}-jet of Ω by adding a term $p^{-1}C_{p^h,p}$.

10.19 <u>Therefore there is a "natural choice" of the parameter</u> $X_{p^{h+1}}$, i.e. $C_{p^h,p} = X_{p^{h+1}}$, $p \in P$, $h \in \underline{\underline{N}}$, <u>which defines completely</u> $A_{p^{h+1}}$.

We have, for $p \in P$

$$A_p = C_{1,p} = X_p$$

$$A_{p^2} = p\, C_{p,p} + C^p_{1,p} A_p = pX_{p^2} + X^{p+1}_p .$$

etc. (see [8]).

10.20 <u>Second case</u>: q <u>is not a prime power</u>. Assume that q has the prime divisors $p_1,\ldots,p_\alpha,\ldots,p_r$, $q = p_\alpha n_\alpha$. Then there are r equations (10.14), namely

10.21 $$A_q = n_\alpha C_{n_\alpha,p_\alpha} + \Sigma_{d|n_\alpha,d<n_\alpha}\, d\, C^{n_\alpha/d}_{d,p_\alpha} A_{n_\alpha/d} \quad (1\leq\alpha\leq r) .$$

So we have r congruences of the form

10.22 $$A_q \equiv \varrho_\alpha \mod. n_\alpha , \quad \text{where } \varrho_\alpha \in \mathfrak{R}_{q-1} .$$

<u>By the extension theorem they must have a solution in</u>

$\mathfrak{R}_{q-1}$, say $a_q \in \mathfrak{R}_{q-1}$, with corresponding values $c_{n_\alpha, p_\alpha} \in \mathfrak{R}_{q-1}$ verifying (10.21). The congruences (10.21) define A_q modulo q (the lcm of the n_α) in $\mathfrak{R}_q$, in other words, we may and shall put

10.23 $$A_q = a_q + qX_q \ , \ C_{n_\alpha, p_\alpha} = c_{n_\alpha, p_\alpha} + p_\alpha X_q \ , \ .$$

This shows, by induction, that $\mathfrak{R}_q$ <u>is generated by the</u> $C_{n,p}$, $np \leq q$, <u>for any</u> q , moreover that $\eta_q = 1$ in equation (III.5.17).

10.24 There remains to <u>choose</u> the a_q . Let us say that a choice is "<u>economical</u>" iff as many monomials as possible in the expansion of $a_q \in \underline{\underline{Z}}[X_2,\dots,X_{q-1}]$ vanish. For example, if $q = pp'$ (two distinct primes), we may take

$$a_{pp'} = \lambda p \, X_p^{p'} X_{p'} + \mu p' \, X_{p'}^p \, X_{p'} \ ,$$

where $\lambda, \mu \in \underline{\underline{Z}}$ verify $\lambda p + \mu p' = 1$. I don't know what a "natural choice" of λ and μ should be.

10.25 Henceforth, we assume that the choice of a_q in (10.23) is economical. <u>We give every variable</u> X_n <u>the weight</u> $n - 1$. <u>Then</u> A_n <u>is isobaric of weight</u> $n - 1$, $C_{n,p}$ <u>is isobaric of weight</u> $np - 1$, <u>and the coefficient</u> $B_{i,j}$ <u>of the universal group law</u> F <u>is isobaric of weight</u> $i + j - 1$. That is proved by induction, using equations (10.14), and remarking that the group law $T^{-1}F(Tx,Ty)$, where T is a new variable, has coefficients $T^{i+j-1} B_{i,j}$, that A_n is replaced by $T^{n-1}A_n$ (see 10.16), and therefore $C_{n,p}$ by $T^{np-1}C_{n,p}$.

10.26 <u>Let</u> S <u>be any set of primes, and put</u> $X_n = 0$ <u>for</u> $n \notin \underline{\underline{S}}$. <u>Then</u> $A_n = 0$ <u>for</u> $n \notin \underline{\underline{S}}$ <u>and</u> $C_{n,p} = 0$ <u>for</u> $np \notin \underline{\underline{S}}$. That follows from (10.14) and our "economy principle". By (10.11), it means that we obtain the universal S-typical group law by just putting $X_n = 0$ for $n \notin \underline{\underline{S}}$, and letting the parameters X_n with $n \in \underline{\underline{S}}$ free as before.

10.27 Let now I be any set. To parametrize the curvilinear group laws on the model $D^{(I)}$, we need parameters $X_{n,i,j}$, where n is an integer ≥ 2 as before, and $i,j \in I$. We consider the $(X_{n,i})_{j \in I}$ as vector variables in a free module with basis indexed by I. Strictly speaking, this makes no sense when I is infinite, therefore we have to put the following definition.

10.28 A specialization $X_{n,i,j} \mapsto x_{n,i,j}$ of the variables $X_{n,i,j}$ in elements $x_{n,i,j}$ of a basic ring K is said to be admissible iff, for any $n \geq 2$ and $i \in I$, almost all $x_{n,i,j}$ vanish.

10.29 Theorem. Let each variable $X_{n,i,j}$ $(n \geq 2,\ i,j \in I)$ receive the weight $n - 1$. Then there are formal series with integral coefficients, $C_{n,p,i,j}$ $(n \in \underline{\underline{N}},\ p \in P,\ i,j \in I)$, isobaric of respective weights $np - 1$ in the $X_{n,i,j}$, such that, for any admissible specialization $X_{n,i,j} \mapsto x_{n,i,j}$ in K, the series $C_{n,p,i,j}(x)$ become polynomials in the $x_{n,i,j}$, so that

10.30 $(C_{n,p,i,j}(x))_{j \in I} \in K^{(I)}$ for any given n,p,i.

If n is a power of p, then

10.31 $$C_{n,p,i,j} = X_{np,i,j}$$

10.32 If n is not a power of p, then

$$C_{n,p,i,j} - p\, X_{np,i,j}$$

is a series in the $X_{n',i',j'}$, where $n' | np$, $n' < np$.

For any set of primes S and any basic ring K, the

S-typical curvilinear group laws on the model $D^{(I)}$ are in one-to-one correspondance with the admissible specializations of the variables $X_{n,p,i,j}$ in K $(X_{n,i,j} \mapsto x_{n,i,j})$ such that $x_{n,p,i,j} = 0$ for $n \notin \underline{\underline{S}}$, the correspondance being expressed by the structural equations

10.33 $$F_p \cdot \gamma_i = \Sigma_{n \in \underline{\underline{S}}, j \in I} V_n[C_{n,p,i,j}(x)] \cdot \gamma_j$$

$(p \in S,\ i \in I)$, the γ_i being the canonical curves of the model $D^{(I)}$.

10.34 For any $q \geqslant 2$, the $x_{n,i,j}$ with $n \leqslant q$, the $C_{n,p,i,j}(x)$ with $np \leqslant q$, and the coefficients of the group law of total degree $\leqslant q$ generate the same subring of K .

The series $C_{n,p,i,j}$ are constructed by induction on np, together with the vector series $A_{i,n}$ in the variables $X_{n,i,j}$, such that $A_{i,1} = e_i$ (the unit vector of index i), and

10.35 $$A_{i,pn} = \Sigma_{d|n, j \in I}\ d\ C^{n/d}_{d,p,i,j}\ A_{j,n/d} ,$$

for any $i \in I$, $p \in P$, $n \in \underline{\underline{P}}$. "Economical" choices are defined as in the one-dimensional case (see 10.24).

11. A digression concerning derivatives

11.1 In (I.6), we introduced the tangent space $\mathfrak{T}V$ of a formal variety V as the module of tangent vectors $\mathfrak{T}\gamma$ to curves $\gamma \in \mathfrak{C}(V)$ at the origin. We could write

11.2 $$\mathfrak{T}\gamma = \frac{d}{dt}\ \gamma(t)\Big|_{t=o} .$$

It is possible, and sometimes desirable, to extend to formal varieties all the properties of analytic manifold that are "local" (at one point), but not necessarily "punctual".

Namely it makes sense to speak of a tangent vector to a formal variety V at a point $x \in V(A)$, $A \in \underline{\underline{nil}}(K)$.

11.3 Let us consider first a morphism of <u>formal modules</u>, $f : L^+ \to M^+$, where L and M are fre K-modules. Then the morphism $(x,y) \mapsto f(x+y)$, $L^+ \times L^+ \to M^+$, splits as

11.4 $$f(x+y) = \Sigma_{m,n \in \underline{\underline{N}}, m+n>o} f_{m,n}(x,y) ,$$

where $f_{m,n}$ is bihomogeneous of bidegree (m,n): see (I.7.3).

For any K-algebra $A \in \underline{\underline{nil}}(K)$, let us denote by A_ε the K-algebra $A \otimes \varepsilon A$, obtained by adjoining to A an element ε verifying $\varepsilon^2 = 0$. Then, for any $x, \xi \in L^+(A)$, i.e. $x, \xi \in A \otimes_K L$, we have $x + \varepsilon \xi \in L^+(A_\varepsilon)$, and formula (11.4) enables us to compute $f(x+\varepsilon\xi) \in M^+(A_\varepsilon)$. We have

11.5 $$f(x+\varepsilon\xi) = \Sigma_{m \in \underline{\underline{P}}} f_{m,o}(x) + \varepsilon \Sigma_{m \in \underline{\underline{N}}} f_{m,1}(x,\xi) ,$$

because all other terms vanish, due to the condition $\varepsilon^2 = 0$.

The <u>derivative of</u> f, denoted by Df, is defined by the formula

11.6 $$f(x+\varepsilon\xi) = f(x) + \varepsilon\, Df(x)\cdot\xi , \quad \varepsilon^2 = 0 ,$$

which is just a way to rewrite (11.5); namely,

11.7 $$Df(x)\cdot\xi = \Sigma_{m \in \underline{\underline{N}}} f_{m,1}(x,\xi) \in M^+(A) .$$

11.8 For any given $A \in \underline{\underline{nil}}(K)$ and $x \in L^+(A)$, <u>we consider</u> $Df(x)$ <u>as a</u> $(K \oplus A)$-<u>linear map of</u> $L^+(A) = A \otimes_K L$ <u>into</u> $M^+(A) = A \otimes_K M$. When $x = 0$, we may identify $Df(0)$ with $\mathfrak{T}f$. Generally, we say that ξ <u>is a tangent vector to</u> L^+ <u>at the point</u> x, <u>which is mapped on the tangent vector</u> $Df(x)\cdot\xi$ <u>to</u> M^+ <u>at the point</u> $f(x)$. Equivalently, together with the map $f_A\colon L^+(A) \to M^+(A)$ we consider the map $Tf_A = f_{A_\varepsilon} : L^+(A_\varepsilon) \to M^+(A_\varepsilon)$.

11.9 The last formulation makes sense for formal varieties, not only formal modules. Together with a formal variety V we consider the "formal tangent bundle" TV, defined as the functor $A \mapsto V(A_\varepsilon)$; the "formal bundle morphism" $TV \to V$ is defined by the K-algebra homomorphism $A_\varepsilon \to A$ (mapping ε on 0). For any morphism of formal varieties, $f : V \to W$, we have the morphism $Tf : TV \to TW$ defined by $Tf_A = f_{A_\varepsilon}$ for any $A \in \underline{\underline{\text{nil}}}(K)$, and the commutative diagram

11.10
$$\begin{array}{ccc} TV & \xrightarrow{Tf} & TW \\ \downarrow & & \downarrow \\ V & \xrightarrow{f} & W \end{array}$$

11.11 Any choice of a formal module structure on V defines a splitting of the formal tangent bundle TV as a product: $TV \simeq V \times \mathfrak{t}V^+$, but the splitting depends on the choice; in other words, "formal constant vector fields" are not intrinsically defined. There is no derivative that could be defined independently of the morphism.

11.12 In the case of a formal group G (not necessarily commutative) the formal tangent bundle TG splits: in differential geometry, the analogue is to consider a Lie group as a left (or a right) homogenous space over itself, defining thereby the "constant" vector fields by the property of being invariant by left (or right) translations. In the language of formal varieties, the natural inclusion $A \to A_\varepsilon = A \otimes \varepsilon A$ gives the $G(A) \to G(A_\varepsilon)$, the 0 section of the formal tangent bundle, and the natural inclusion $\varepsilon A \to A \otimes \varepsilon A$ gives

11.13
$$\mathfrak{t}G^+(A) \xrightarrow{\sim} G(\varepsilon A) \to G(A_\varepsilon) .$$

(identification of $\mathfrak{L}G$ with the tangent space to the origin).

Any $v \in G(A_\varepsilon)$ may be written, in one and only one way, as $v = \mu_{A_\varepsilon}(x,\xi)$, where $x \in G(A)$, $\xi \in \mathfrak{L}G^+(A) = A \otimes_K \mathfrak{L}G$ are identified with their natural images in $G(A_\varepsilon)$, while $\mu : G \times G \to G$ denotes the group morphism. A much more suggestive notation is

11.14 $$v = x +_G \varepsilon \xi$$

11.15 Definition. Let $f : G \to G'$ be a morphism of (non necessarily commutative) formal groups. Then the reduced (left) derivative of f, denoted by $\breve{D}f$, is defined by the formula

11.16 $$f(x +_G \varepsilon \xi) = f(x) +_{G'} \varepsilon \breve{D}f(x)\cdot\xi \quad , \quad \varepsilon^2 = 0 ,$$

where we have omitted the subscript A, as we did in (11.6).

For a systematic development of the calculus with reduced derivatives, see [11]. Here, let us be content to state that the reduced derivative of a curve $\gamma : D \to G$ in a formal group is defined as in (11.16) where G is replaced by $\underline{\underline{G}}_a$ (or K^+) and G' by G.

11.17 $$\gamma(t+\varepsilon\theta) = \gamma(t) +_G \varepsilon \breve{D}\gamma(t)\theta \quad , \quad \varepsilon^2 = 0 .$$

Then, as we said, the definition (7.2) has be proved, by using the lift theorem (see II.3.8).

11.18 Proposition. Let G be a formal group defined on a formal module L^+ by the group morphism $\mu : L^+ \times L^+ \to L^+$. Then the reduced derivative $\breve{D}\gamma$ of a curve $\gamma \in \mathfrak{C}(G)$ is defined by

11.19 $$\gamma'(t) = D_2\mu(\gamma(t),0)\cdot\breve{D}\gamma(t) \quad ,$$

where γ' denotes the ordinary derivative of γ and $D_2\mu$ the ordinary second partial derivative of μ.

CHAPTER VI

ON FORMAL GROUPS IN CHARACTERISTIC p

1. Notations for the local case

1.1 In this chapter, and in the following one, p will denote a fixed prime, and we shall consider only basic rings where all primes except p will be invertible. Therefore, by the reduction theorem (see IV.8.1), we shall be in the "local case".

1.2 We shall write "p-typical" and "$\mathbb{C}_p(G)$" instead of "{p}-typical" and "$\mathbb{C}_{\{p\}}(G)$".

We shall write simply F,V, instead of F_p, V_p, and also, over a basic ring K ,

1.3
$$\begin{cases} E \text{ or } E(K) \text{ instead of } \mathrm{Cart}_{\{p\}}(K) & \text{(see IV.2)}, \\ E_+ \text{ or } E_+(K) \text{ instead of } \mathrm{Cart}_{\{p\}}(K)_+ & \text{(see V.5.11)}, \\ W \text{ or } W(K) \text{ instead of } W_{\{p\}}(K) & \text{(see IV.4)}. \end{cases}$$

So W(K) will denote the ring of "classical" Witt vectors. Concerning their origin, and a fascinating period in the development of arithmetical algebra, we refer the reader to Witt's original paper [16].

1.4 Some changes must be introduced in our notations, because we want to write a (classical) Witt vector $\xi \in W(K)$ as $(a_o, a_1, \ldots, a_n, \ldots)$, not as $(a_1, a_p, \ldots, a_{p^n}, \ldots)$. More generally, functions which, according to our general conventions, had to take their values in $p^{\underline{\underline{Z}}}$, will now be allowed to take them simply in $\underline{\underline{Z}}$, if possible. We replace p^h by h , when it makes sense and brings a simplification.

For instance, the functorial ring homomorphisms, w_n: W(K) → K are now defined for $n \in \underline{\underline{N}}$ by the formulas

1.5 $$w_n(\xi) = \Sigma_{o\leq i\leq n}\, p^i a_i^{p^{n-i}} \quad , \quad \xi = (a_o, a_1, \ldots) \in W(K) \ ,$$

but the weight of the component (or coordinate) a_n is p^n as before, not n (see IV.4.6).

Any element $x \in E(K)$ has a unique expansion as

1.6 $$x = \Sigma_{m,n\in\underline{\underline{N}}}\, V^m[a_{m,n}]F^n \quad , \quad a_{m,n} \in K \ ,$$

with the finiteness condition that $(a_{m,n})_{n\in\underline{\underline{N}}} \in K^{(\underline{\underline{N}})}$ for any m ; its order, denoted henceforward by $\mathrm{ord}(x)$ is the smallest m for which there is an $a_{m,n} \neq 0$. We obtain a complete description of the functor in topological rings E , by adding the following rules (see IV.2), where $a,b \in K$:

1.7 $$\begin{cases} [1_K] = 1_E \ , & [ab] = [a][b] \ , \quad \mathrm{ord}([a+b] - [a] - [b]) > 0 \\ FV = p1_K \ , & [a]V = V[a^p] \ , \quad F[a] = [a^p]F \ . \end{cases}$$

We shall often identify a Witt vector $(a_n)_{n\in\underline{\underline{N}}} = \xi$ with its natural image in E ,

1.8 $$\mathrm{op}(\xi) = \Sigma_{n\in\underline{\underline{N}}}\, V^n[a_n]F^n \quad ,$$

but we must be careful to distinguish $F\xi \in E$ from the transformed of ξ by F acting on W ; we hope to prevent confusion by writing the later as ξ^F , so that we have the following commutation rules

1.9 $$\begin{cases} \xi V = V\xi^F \ , \ F\xi = \xi^F F \quad (\xi \in W) \ ; \quad [a]^F = [a^p] \quad (a \in K) \ . \\ w_n(\xi^F) = w_{n+1}(\xi) \qquad \text{(see IV.4.22)} \ . \end{cases}$$

1.10 The subring E_+ of E is defined by adding in (1.6) the condition $a_{m,n} = 0$ when $m < n$. We may identify E_+ with the ring of twisted formal series $\xi_o + V\xi_1 + \ldots + V^n\xi_n + \ldots = e$, with the commutation rules $\xi V^n = V^n \xi^{F^n}$, but the order functions are inherited from E . More precisely, for $\xi = (a_n)_{n\in\underline{\underline{N}}} \in W(K)$, $\mathrm{ord}(\xi)$ is the smallest n with $a_n \neq 0$, while, for

$$e = \Sigma_{n\in\underline{\underline{N}}} V^n \xi_n \in E_+ \quad (\xi_n \in W) ,$$

$$\mathrm{ord}(e) = \min_{n\in\underline{\underline{N}}} (n+\mathrm{ord}(\xi_n)) .$$

The definitions of uniform and reduced E-modules are now much simpler than in (IV.5).

1.11 A left E-module C is said to be uniform iff it is Hausdorff and complete for its V-adic topology. In other words, the natural map $C \to \varprojlim C/V^nC$ is a bijection.

More explicitely

1.12 $$\cap_{n\in\underline{\underline{N}}} V^nC = 0 ,$$

and any set of pairwise compatible congruences $\gamma \equiv \gamma_n$ modulo V^nC ($n \in \underline{\underline{N}}$ and $\gamma_{n+1} \equiv \gamma_n$ mod. V^nC) has a solution γ in C.

A reduced E-module C is a uniform E-module such that

1.13 $V\cdot\gamma = 0$ implies $\gamma = 0$ for any $\gamma \in C$,

and that C/VC is a free K-module.

Indeed, condition (1.13) implies that the maps $V^n: C/VC \to V^nC/V^{n+1}C$ are injective for any $n \in \underline{\underline{N}}$. We express (1.13) by saying that C is V-torsion-free, or that V acts injectively on C.

The matrix representation κ' of E (see IV.3.11) is now indexed by $\underline{\underline{N}}$, not by $p^{\underline{\underline{N}}}$; we write, with the matrix units $(\underline{\underline{e}}_{m,n})_{m,n\in\underline{\underline{N}}}$,

1.14
$$\begin{cases} \kappa'(F) = \Sigma_{n\in\underline{\underline{N}}} \underline{\underline{e}}_{n,n+1} , \\ \kappa'(V) = \Sigma_{n\in\underline{\underline{N}}} p\, \underline{\underline{e}}_{n+1,n} , \\ \kappa'([c]) = \Sigma_{n\in\underline{\underline{N}}} c^{p^n} \underline{\underline{e}}_{n,n} , \quad c \in K . \end{cases}$$

For $x \in E$, $x = \Sigma_{m,n} V^m[x_{m,n}]F^n$, $x_{m,n} \in K$, we have

1.15 $$\kappa'(x) = \Sigma_{m,n,i\in\underline{\underline{N}}} p^m x^{p^i}_{m,n} \underline{\underline{e}}_{m+i,n+i} ,$$

or equivalently

1.16 $$\kappa'(x)_{m,n} = \Sigma_{o\leq i\leq \min(m,n)} p^{m-i} x^{p^i}_{m-i,n-i} .$$

1.17 If M is a K-module, then $M_p[[t]]$ denotes the additive group of power series

1.18 $$f(t) = \Sigma_{n\in\underline{\underline{N}}} t^{p^n-1} a_n , \quad a_n \in M \quad \text{(see V.7.8)}$$

and the left action of E on $M_p[[t]]$ is defined by the formula

1.19 $$(x\cdot f)(t) = \Sigma_{n\in\underline{\underline{N}}} t^{p^n-1} b_n ,$$

where f is as in (1.18), $x \in E$ as in (1.15), and

1.20 $$b_n = \Sigma_{m\in\underline{\underline{N}}} \kappa'(x)_{n,m} a_m = \Sigma_{o\leq i\leq n, m\in\underline{\underline{N}}} p^i x^{p^{n-i}}_{i,m} a_{m+n-i} .$$

1.21 For a formal group G over K, the map $\check{D} : \mathfrak{C}_p(G) \to \mathfrak{T}G_p[[t]]$ is E-linear (see V.7.11). When G is given by a presentation

1.22 $$F\cdot\gamma_i = \Sigma_{m,n\in\underline{\underline{N}},j\in I} V^{m+n}[c_{m,n,i,j}]F^n\cdot\gamma_j , \quad i \in I$$

(see V.5.15), then the reduced derivatives

1.23 $$D\gamma_i(t) = \Sigma_{n\in\underline{\underline{N}}} t^{p^n-1} a_{i,n} , \quad i \in I, \ a_{i,n} \in \mathfrak{T}G \quad \text{(see V.7.14)}$$

are defined by the formulas (see V.7.18)

1.24 $$a_{i,n+1} = \Sigma_{\alpha,\beta\in\underline{\underline{N}},\alpha+\beta\leq n,j\in J} p^{\alpha+\beta} c^{p^{n-\alpha-\beta}}_{\alpha,\beta,i,j} a_{j,n-\alpha} , \quad i \in I, \ n \in \underline{\underline{N}} .$$

1.25 The vectors $a_{i,o} = \mathfrak{T}\gamma_i$ are a basis of $\mathfrak{T}G$ over K. The <u>structural constants of</u> G (see V.5.17), $c_{n,i,j}$, $n \in \underline{\underline{N}}$, $i,j \in I$, are the coefficients $c_{n,o,i,j}$ in (1.22), <u>assuming that</u> $m > 0$ <u>implies</u> $c_{n,m,i,j} = 0$. Then formulas (1.24) become

1.26 $$a_{i,n+1} = \Sigma_{o\leq\alpha\leq n,j\in I} p^\alpha c^{p^{n-a}}_{\alpha,i,j} a_{j,n-\alpha} , \quad i \in I, \ n \in \underline{\underline{N}} .$$

2. The special features of characteristic p

2.1 *From now on (in this chapter) we assume that the basic rings* K *have characteristic* p . This is expressed by any one of the three following *equivalent properties*:

2.2 $$p1_K = 0 \;;$$

2.3 *the map* $x \mapsto x^p$ *is an endomorphism of* K ;

2.4 $$VF = FV = p \quad \text{in} \quad E(K) \quad \text{(see IV.4.16)} .$$

2.5 *Proposition. In any* E-*module* C , *the* p-*adic topology is finer than the* V-*adic topology, and* C *is complete for its* p-*adic topology when it is so for its* V-*adic topology* (the converse may be false).

Let C *be a left* E_+-*module, such that* $C \simeq \varprojlim C/V^nC$ (see 1.11), *and that* V *acts injectively on* C . *Then* C *has an* E-*module structure (extending that of* E_+-*module) iff*

2.6 $$pC \subset VC .$$

In that case, $F\cdot\gamma$ *is computed for any* $\gamma \in C$ *by the formula* $V\cdot(F\cdot\gamma) = p\gamma$, *or* $F\cdot\gamma = V^{-1}p\gamma$.

Proof. The first assertions follow from the formula

2.7 $$p^n = V^nF^n = F^nV^n \quad , \quad n \in \underline{\underline{N}} .$$

Condition (2.6) is obviously necessary for an E_+-module to have an E-module structure. Together with the injectivity of V, it enables one to define an additive operator F by $VF\gamma = p\gamma$. Obviously, F commutes to V , and the formula $F[a] = [a^p]F$ is equivalent to $VF[a] = V[a^p]F$ and, as $V[a^p] = [a]V$, it reduces to $p[a] = [a]p$. Then the condition $C \simeq \varprojlim C/V^nC$ suffices for the complete ring E to act on C , for the existence of the E-module structure on C follow from (1.7).

2.8 Proposition. For any $m,n,i,j \in \underline{\underline{N}}$ and $a,b \in K$, we have

2.9 $$V^m[a]F^n V^i[b]F^j = V^{m+i}[a^{p^i} b^{p^n}]F^{n+j} .$$

The order functions on $E = E(K)$ and also, by restriction, on E_+ and on W, satisfy the relation

2.10 $$\mathrm{ord}(xy) \geq \mathrm{ord}(x) + \mathrm{ord}(y) .$$

The ring endomorphism $\xi \mapsto \xi^F$ of W may now be extended to the whole of E, so that the following rules hold:

2.11 $$xV = Vx^F , \ Fx = x^F F , \ x \in E .$$

2.12 The ring endomorphism $x \mapsto x^F$ of E is the functorial morphism associated to the endomorphism $a \mapsto a^p$ of K. An element $x = \Sigma_{m,n \in \underline{\underline{N}}} V^m[a_{m,n}]F^n$ (see 1.6) becomes $x^F = \Sigma_{m,n \in \underline{\underline{N}}} V^m[a^p_{m,n}]F^n$. For a Witt vector $\xi = (a_n)_{n \in \underline{\underline{N}}}$, we have $\xi^F = (a_n^p)_{n \in \underline{\underline{N}}}$.

2.13 A Witt vector $\xi = (a_o,a_1,\ldots)$ is invertible in W iff a_o is invertible in K; an element $e = \xi_o + V\xi_1 + \ldots \in E_+$ is invertible in E_+ iff ξ_o is invertible in W.

Proof. Relation (2.9) follow from (2.4) and (1.7); it implies (2.10) and (2.11), with the given formula for x^F, $x \in E$. By (IV.2.5), $x \mapsto x^F$ in E is associated to $a \mapsto a^p$ in K.

If $\xi = (a_o,a_1,\ldots)$ is invertible in W or $e = \xi_o + V\xi_1 + \ldots$ is invertible in E_+, then, by (2.10), (2.9), a_o (resp. ξ_o) has to be invertible in K (resp. in W). Conversely, if this condition is verified, we have $\xi[a_o^{-1}] = (1,b_1,\ldots)$ (resp. $e\xi_o^{-1} = 1 + V\eta_1 + \ldots$). If x is any element of E such that $\mathrm{ord}(x-1) > 0$, we have

2.14 $$x^{-1} = \Sigma_{n \in \underline{\underline{N}}} (1-x)^n .$$

The general construction relative to a "change of rings"

$\varphi : K \to K'$ (see II.2.11) will be applied, in the next proposition, to the endomorphism $\varphi : a \mapsto a^p$ of K. We must appreciate that φ is not yet an automorphism so that there is still a distinction between domain and range of φ.

2.15 Proposition. Let G be a formal group over K and $\mathfrak{C}_p(G)$ its module of p-typical curves, defined by the presentation

2.16 $$F\cdot\gamma_i = \Sigma_{j\in I}\, x_{i,j}\cdot\gamma_j \quad , \quad i \in I,\ x_{i,j} \in E_+ \quad \text{(see V.5.15)} .$$

Let G' be the formal group over K constructed from G with the "change of rings" $a \mapsto a^p$, and $\gamma \mapsto \gamma'$ the natural map $\mathfrak{C}_p(G) \to \mathfrak{C}_p(G')$, so that

2.17 $$F\cdot\gamma_i' = \Sigma_{j\in I}\, x_{i,j}^F\cdot\gamma_j'$$

is the corresponding presentation of $\mathfrak{C}_p(G')$.

Then there are functorial formal group homomorphisms over K,

2.18 $G \to G'$, mapping $\gamma \in \mathfrak{C}_p(G)$ on $V\cdot\gamma' \in \mathfrak{C}_p(G')$,

2.19 $G' \to G$, mapping $\gamma' \in \mathfrak{C}_p(G')$ on $F\cdot\gamma \in \mathfrak{C}_p(G)$,

and their composed homomorphisms are the multiplication by p in G and in G'.

Proof. Formula (2.17) follows from (IV.1.6) and (2.12). Then it follows from (2.11) that the curves $V.\gamma_i' \in \mathfrak{C}_p(G')$ verify the relations

2.20 $$F\cdot V\cdot\gamma_i' = \Sigma_{j\in I}\, x_{i,j}\ V\cdot\gamma_j' ,$$

and similarly that the curves $F\gamma_i$ in $\mathfrak{C}_p(G)$ verify

2.21 $$F\cdot F\cdot\gamma_i = \Sigma_{j\in I}\, x_{i,j}^F\ F\cdot\gamma_j .$$

By (V.5.4), it proves the existence of the formal group homomorphisms (2.18), (2.19), where γ, γ' are replaced respectively by γ_i, γ_i' ($i \in I$). But then one checks that these formulas hold for any p-typical curve, and more generally for any curve $\gamma \in \mathbb{C}(G)$ or γ' in the image of $\mathbb{C}(G)$ in $\mathbb{C}(G')$. The property of the composed homomorphisms results from (2.4).

3. Fields and perfect fields

3.1 From now on (in this chapter), we assume that K is a field of characteristic p .

Then every K-module is free and, by (1.12)

3.2 a reduced E-module is a V-torsion-free E-module, complete for its V-adic topology. Equivalently, it is an E_+-module, satisfying the preceding conditions, and also $VC \supset pC$ (see 2.5).

3.3 Proposition. Over a field K of characteristic p , both W(K) and $E_+(K)$ are local rings, where invertible (resp. non-invertible) elements have order 0 (resp. > 0). Their residual fields are K .

That follows from (2.13)

3.4 Proposition. Any closed submodule C' of a reduced E-module C, together with its own V-adic topology, is reduced.

That follows from (3.2), as explicited in (1.12).

3.5 In the additive category of formal groups over a field, morphisms have kernels, but generally not cokernels; however, kernels have cokernels. That is expressed by the following proposition

3.6 Proposition. Let $f : G \to G'$ be a morphism of formal groups. Then there are formal groups N , H , together with morphisms $g : N \to G$, $h : G \to H$, such that the following sequences are exact

3.7
$$0 \to \mathfrak{C}_p(N) \to \mathfrak{C}_p(G) \to \mathfrak{C}_p(G') ,$$
$$0 \to \mathfrak{L}N \to \mathfrak{L}G ;$$

3.8
$$0 \to \mathfrak{C}_p(N) \to \mathfrak{C}_p(G) \to \mathfrak{C}_p(H) \to 0$$
$$0 \to \mathfrak{L}N \to \mathfrak{L}G \to \mathfrak{L}H \to 0 .$$

3.9 In other words, the kernels of formal group homomorphisms, $f : G \to G'$, may now be identified with <u>embedded subgroups of</u> G (see V.4.13); they are in one-to-one correspondence with the closed E-<u>submodules</u> C' <u>of</u> $C = \mathfrak{C}_p(G)$ <u>which are</u> V-<u>divisible in</u> C , <u>i.e. such that</u> $\gamma \in C$, $V.\gamma \in C'$ <u>imply</u> $\gamma \in C'$ (<u>or, equivalently</u>, C/C' <u>is</u> V-<u>torsion free</u>). Then C/C' is reduced, by (3.2), and corresponds to a <u>factor formal group of</u> G (see V.4.17)

3.10 If C is a reduced E-module, then VC is reduced, but its dimension is that of C multiplied by the algebraic degree $(K:K^p)$, where K^p denotes the subfield of p-th powers in K , because $[c]V\cdot\gamma = V[c^p]\cdot\gamma$ for $\gamma \in C$, $c \in K$. That is the reason why we do not speak of non-embedded subgroups in a formal group when the basic field is not <u>perfect</u>.

3.11 <u>From now on, we assume that</u> K <u>is a perfect field of characteristic</u> p , i.e. $K^p = K$ and $a \mapsto a^p$ is an <u>automorphism</u> of K , whose inverse is denoted by $a \mapsto a^{p^{-1}}$.

Then $\xi \mapsto \xi^F$ is an <u>automorphism</u> of $W(K)$ (see 1.9) and $x \mapsto x^F$ is an <u>automorphism</u> of $E(K)$, or of $E_+(K)$ (see 2.11).

In the general expansion (1.6) of an element of E , it is no more necessary to write V on the left and F on the right, because

3.12
$$V^m[a]F^n = F^n[a^{p^{-m-n}}]V^m , \quad m,n \in \underline{\underline{N}} , \quad a \in K .$$

3.13 <u>Proposition</u>. <u>Over a perfect field</u> K , <u>a Witt vector</u>

$\xi = (a_n)_{n \in \underline{\underline{N}}}$ can be written equivalently as

3.14 $$\xi = \Sigma_{n \in \underline{\underline{N}}} p^n [a_n^{p^{-n}}] .$$

The order of ξ is its p-adic order: $\mathrm{ord}(\xi) \geq n$ iff $p^n | \xi$ (for any $n \in \underline{\underline{N}}$). The radical of W, i.e. its maximum ideal (or the kernel of $w_o : W \to K$) is pW. So W is a complete p-adic local ring, with residual field K, which is "absolutely unramified" (i.e. $\mathrm{ord}(p) = 1$).

3.15 Let C be an E-module. Then the image FC is an E-submodule of C.

3.16 For any E-submodule C' of C, we define the V-divided module of C' in C as the closure of the set of all $\gamma \in C$ such that $V^n \cdot \gamma \in C'$ for some $n \in \underline{\underline{N}}$; it is an E-submodule of C, which we denote by $\mathrm{div}_C C'$.

All those assertions follow from (3.12), and would not hold if K was not perfect.

3.17 Proposition. Let C' be a closed submodule of a reduced E-module C. Then there is a V-basis $(\gamma_i)_{i \in I}$ of C, and a partition of the indexing set: $I = I_\infty \cup \cup_{n \in \underline{\underline{N}}} I_n$, with the following property.

An element $\gamma \in C$ has a unique expansion as

3.18 $$\gamma = \Sigma_{m \in \underline{\underline{N}}, i \in I} V^m [a_{m,i}] \cdot \gamma_i \quad , \quad a_{m,i} \in K ,$$

where $(a_{m,i})_{i \in I} \in K^{(I)}$ for any $m \in \underline{\underline{N}}$, and

3.19 $$\gamma \in C' \text{ iff } a_{m,i} = 0 \text{ when } i \in I_n \text{ and } m < n$$

(in particular, $a_{m,i} = 0$ for any m when $i \in I_\infty$).

Proof. We consider on C' the filtration induced by that of C, i.e. the submodules $V^n C \cap C'$ $(n \in \underline{\underline{N}})$, and we put

3.20 $$gr_n C' = (V^n C \cap C')/(V^{n+1} C \cap C') .$$

Every $gr_n C'$ has a natural structure of K-vector space, more precisely of vector subspace of $gr_n C = V^n C/V^{n+1} C$. The multiplication by $c \in K$ in $gr_n C$ corresponds to multiplication by $[c] \in W(K)$ in $V^n C$.

The multiplication by V in C induces semi-linear isomorphisms $V : gr_n C \to gr_{n+1} C$, because K is perfect, and $V\, gr_n C' \subset gr_{n+1} C'$.

Let us choose, for any $n \in \underline{\underline{N}}$, a supplement Γ_n to $V\, gr_{n-1} C'$ in $gr_n C'$, so that

3.21 $$gr_n C' = \Gamma_n \oplus V\, gr_{n-1} C' \qquad (\Gamma_o = gr_o C') .$$

Then we have direct decompositions

3.22 $$gr_n C' = \oplus_{o \leq i \leq n} V^i\, \Gamma_{n-i} \ , \ n \in \underline{\underline{N}} .$$

In order to obtain the $\gamma_i \in C$ for $i \in I_n$, we choose a basis of Γ_n indexed by I_n , we choose representatives of this basis in $V^n C \cap C'$, and we divide them by V^n. The sets I_n are assumed to be disjoint. It follows from (3.22) that the γ_i are linearly independant modulo VC for $i \in \cup_{n \in \underline{\underline{N}}} I_n$. So we can take a V-basis $(\gamma_i)_{i \in I}$ of C by adjoining (eventually) new elements γ_i , where $i \in I_\infty$ (an indexing set, disjoint from the preceding ones), and putting $I = I_\infty \cup \cup_{n \in \underline{\underline{N}}} I_n$.

By construction, the $V^n \gamma_i$, $i \in \cup_{o \leq \alpha \leq n} I_\alpha$, taken modulo $V^{n+1} C \cap C'$, are a basis of Γ_n , so that (3.20) follows readily from (3.22).

3.23 Remarks. The hypothesis that C' is closed is superfluous when C is finite-dimensional; then the $V^n \gamma_i$, $i \in I_n$, $n \in \underline{\underline{N}}$, are a V-basis of C' .

Unless $I_n = \emptyset$ for n large enough, the V-topology of C' is not induced by that of C.

The V-divided module $\mathrm{div}_C C'$ (see 3.16) is generated generally by the γ_i where $i \in \cup_{n \in \underline{\underline{N}}} I_n$.

4. Finite dimension and isogeny

4.1 *From now on we shall consider only formal groups of finite dimension, over a perfect field* K.

If $C = \mathbb{C}_p(G)$ is the E-module of p-typical curves in a formal group G over K, then $gr_o C = C/VC$ is identified with the tangent space $\mathfrak{T}G$ of G, and the new assumption means that

4.2 $$\dim G = \dim_K \mathfrak{T}G = \dim_K C/VC < \infty .$$

4.3 *Proposition*. *Let* $f : G \to G'$ *be a homomorphism of formal groups*, $d = \dim G$, $d' = \dim G'$. *Then there is an integer* $r \leqslant \min(d,d')$, *called the rank of* f, *and a sequence*

$$n_1 \leqslant n_2 \leqslant \ldots \leqslant n_r \quad \text{in } \underline{\underline{N}} ,$$

such that, for suitable curvilinear p-*typical coordinate systems*, $x = (x_1,\ldots,x_d)$ *in* G, $y = (y_1,\ldots,y_{d'})$ *in* G', *the relation* $y = f(x)$ *is expressed by the equations*

4.4 $$\begin{cases} y_i = x_i^{p^{n_i}} , & \text{for } 1 \leqslant i \leqslant r , \\ y_i = 0 , & \text{for } i > r . \end{cases}$$

Proof. That results from proposition (3.6) and (3.17), applied to the image of $\mathbb{C}_p(G)$ in $\mathbb{C}_p(G')$.

Before giving general definitions let us discuss the following example of a 2-dimensional group G defined by the presentation (see V.5.15)

4.5 $$F\cdot\gamma_1 = \gamma_2 \ ,$$

4.6 $$F\cdot\gamma_2 = V\gamma_2 \ .$$

Equation (4.6) shows that there is in G a one-dimensional embedded subgroup G_2 corresponding to the coordinate curve γ_2. More precisely, $\gamma_2(x) + \gamma_2(y) = \gamma_2(\mu(x,y))$, where μ is the group law defined by

4.7 $$\Sigma_{n\in\underline{\underline{N}}}\ p^{-n}\mu(x,y)^{p^{2n}} = \Sigma_{n\in\underline{\underline{N}}}\ p^{-n}(x^{p^{2n}} + y^{p^{2n}}) \qquad \text{(see V.9)}.$$

Equation (4.5) shows that the factor formal group G/G_2 is an additive formal group, but the extension does not split.

In $C = \mathbb{C}_p(G)$, the submodule corresponding to G_2 is $C_2 = FC$. The kernel C' of F on C is generated by $V\gamma_1 - \gamma_2$. With the notation of (3.20), we have $gr_n C_2 = gr_n C'$ for any $n \in \underline{\underline{N}}$, while $C_2 \cap C' = 0$. The sum $C' + C_2$ is direct, but $C' \oplus C_2 \neq C$.

In the coordinate system defined by γ_1 and γ_2, the group morphism of G is written

4.8 $$(x_1,x_2) +_G (y_1,y_2) = (x_1+y_1,\ x_2+_\mu y_2+_\mu f(x_1,y_1))$$

(see V.4.10). The curve $V\gamma_1 - \gamma_2$, i.e. $t \mapsto (t^p, -_\mu t)$ is additive; by writing that dow, we obtain the following formula for the 2-cocycle f:

4.9 $$f(t^p,t'^p) = t +_\mu t' -_\mu (t +_a t') \ ,$$

where "$+_a$" denotes ordinary addition.

4.10 Things would obviously be easier if we could assume that the operation $t \mapsto t^p$ is invertible, i.e. that extraction of a p-th root is to be added to the usual ring operations. That would be absurd in algebras $A \in \underline{\underline{\text{nil}}}(K)$, but we shall presently define an equivalence relation, called <u>isogeny</u>, coarser than isomorphism, on the category of finite dimensional formal groups over the perfect field K . The definitions will first be given for modules, by adjoining an inverse V^{-1} to V ; if γ is a curve, than $V^{-1}\gamma$ can be imagined as $t \mapsto \gamma(t^{p^{-1}})$.

4.11 Definitions. The V-divided ring of $E = E(K)$, denoted by $\tilde{E} = \tilde{E}(K)$, is defined as follow. Its elements are written $V^{-n}x$, where $n \in \underline{\underline{N}}$, $x \in E$, with equality defined by

4.12 $$V^{-m}x = V^{-n}y \quad \text{iff} \quad V^{n}x = V^{m}y ,$$

and operations by

4.13 $$V^{-m}x + V^{-n}y = V^{-(m+n)}(V^{n}x + V^{m}y) ,$$

4.14 $$(V^{-m}x)(V^{-n}y) = V^{-(m+n)}(x^{F^{-n}}y) ,$$

where $x,y \in E$.

The natural embedding of E in $\tilde{E}$ is obtained by identifying $x \in E$ with $V^{-0}x \in \tilde{E}$.

4.15 Similarly, we define the (absolute) V-divided module $\tilde{C}$ of an E-module C by formulas anologous to (4.12), (4.13), with $\gamma,\gamma' \in C$ replacing $x,y \in E$, and the $\tilde{E}$-module structure of $\tilde{C}$ by (4.14) with $y \in E$ replaced by $\gamma \in C$. The natural map $C \to \tilde{C}$, $\gamma \mapsto V^{-0}\gamma$, is an embedding iff C is V-torsion-free.

4.16 One checks easily that, for any exact sequence of E-modules, $0 \to C' \to C \to C'' \to 0$, the corresponding sequence of $\tilde{E}$-modules, $0 \to \tilde{C}' \to \tilde{C} \to \tilde{C}'' \to 0$ is exact. If C' is an E-submodule of C, we may identify its V-divided $\tilde{C}'$ with the $\tilde{E}$-submodule of $\tilde{C}$ generated by C' .

4.17 If, in (4.16), we assume that C is V-torsion free, then any $\tilde{E}$-submodule of $\tilde{C}$ is generated by its intersection with C . If C' is any E-submodule of C , then $\gamma \in C \cap (\tilde{E}C')$

means that there is an $n \in \underline{N}$ such that $V^n \gamma \in C'$. <u>Therefore</u>
<u>there is a one-to-one correspondence between the</u> $\tilde{E}$-<u>submodule of</u>
$\tilde{C}$ <u>and the</u> E-<u>submodules of</u> C <u>which are (relatively)</u> V-<u>divisible.</u>

4.18   <u>Let</u> C <u>be a reduced finite-dimensional</u> E-<u>module</u>
$(\dim_K C/VC < \infty)$. <u>Then</u> $\tilde{C}$ <u>is an</u> $\tilde{E}$-<u>module of finite length (i.e.</u>
<u>it has a Jordan-Hölder sequence).</u>

    Proof. That follows from (4.17) because, if

$$0 = C_o \neq C_1 \neq \cdots \neq C_r = C$$

is an increasing sequence of V-divisible E-submodules in the
finite-dimensional E-module C , we may take a V-basis
$(\gamma_i)_{1 \leq i \leq d}$ of C , such that every C_i admits as a V-basis a
subset of the γ_i , whence $r \leq d$.

4.19   <u>Definitions. We shall say that two (finite-dimensional)</u>
<u>formal groups</u> G, G' <u>over</u> K <u>are isogenic iff the (absolute)</u>
V-<u>divided modules</u> $\tilde{C}$ <u>of</u> $C = \mathfrak{C}_p(G)$ <u>and</u> $\tilde{C}'$ <u>of</u> $C' = \mathfrak{C}_p(G')$
<u>are isomorphic</u> $\tilde{E}$-<u>modules.</u>

4.20   <u>An</u> $\tilde{E}$-<u>module such as</u> $\tilde{C}$ <u>will be called a finite</u> $\tilde{E}$-<u>module,</u>
<u>and an</u> E-<u>submodule of</u> $\tilde{C}$ <u>such as</u> C <u>will be called a "lattice"</u>
<u>in</u> C .

4.21   Assume that, in proposition (4.3), r = rk(f) = dim G = dim G'.
Then the groups G and G' are isogenic, and f is called an
<u>isogeny</u>. Whether a reciprocal f^{-1} of f should be defined
and also called an isogeny (f^{-1} being defined by formulas as
(4.4), but with exponents $n_i \in \underline{Z}$) is left for the reader to de-
cide.

4.22   Anyhow, <u>isogeny is an equivalence relation. An isogeny class</u>
<u>is defined by a finite</u> $\tilde{E}$-<u>module</u> $\tilde{C}$ (up to $\tilde{E}$-isomorphism), <u>and</u>

the isomorphy classes it contains are in one-to-one correspondence with the isomorphy classes of lattices in $\tilde{C}$.

4.23 Proposition. A subset C of a finite $\tilde{E}$-module $\tilde{C}$ is a lattice in $\tilde{C}$ iff it is a finitely generated E-submodule of $\tilde{C}$ such that $\tilde{C}/C$ is a V-torsion module (i.e. for any $\gamma \in \tilde{C}$ there is an $n \in \mathbb{N}$ with $V^n\gamma \in C$).

To any lattice C in $\tilde{C}$ there corresponds an order function on $\tilde{C}$, $\mathrm{ord}_C : \tilde{C} \to \mathbb{Z} \cup \{+\infty\}$, defined by

4.24 $$\mathrm{ord}_C(\gamma) \geq n \quad \text{iff} \quad \gamma \in V^n C \quad \text{for any} \quad \gamma \in \tilde{C},\ n \in \mathbb{Z} .$$

It is the unique extension of the V-order on C satisfying

4.25 $$\mathrm{ord}_C(V^n\gamma) = n + \mathrm{ord}_C(\gamma), \quad \text{for any} \quad \gamma \in \tilde{C},\ n \in \mathbb{Z} .$$

If C and C_1 are any two lattices in $\tilde{C}$, so are their intersection $C \cap C_1$ and their sum $C + C_1$, and there is an $m \in \mathbb{N}$ such that

4.26 $$|\mathrm{ord}_C(\gamma) - \mathrm{ord}_{C_1}(\gamma)| \leq m \quad \text{for any} \quad \gamma \in \tilde{C} .$$

There is a function, denoted by $\lg(C,C_1)$, which is defined for any pair of lattices in $\tilde{C}$ by the following properties:

4.27 if $C_1 \subset C$, then $\lg(C,C_1)$ is the (Jordan-Hölder) length of the E_+-module C/C_1 ; for any three lattices C, C_1, C_2 in $\tilde{C}$,

4.28 $$\lg(C,C_1) + \lg(C_1,C_2) = \lg(C,C_2) .$$

Proof. The first assertions are clear. In order to prove that $C \cap C_1$ is finitely generated, one uses the more general

4.29 property that a lattice is a noetherian E-module, which follows from proposition (3.17) and remark (3.19). If $C_1 \subset C$ are two lattices in $\tilde{C}$, then there is a V-basis $(\gamma_i)_{1 \leq i \leq d}$ in C such that $(V^{n_i}\gamma_i)_{1 \leq i \leq d}$ is a V-basis in C_1 , where $n_1, \ldots, n_d$ is a sequence in $\underline{\underline{N}}$. It implies that <u>the</u> E_+-<u>length of</u> C/C_1 <u>is</u>

4.30 $$lg(C,C_1) = \Sigma_{1\leq i\leq d} \; n_i \quad .$$

<u>The Jordan-Hölder factors of</u> C/C_1 <u>are all isomorphic to</u> K (<u>with trivial action of</u> V). <u>The</u> E-<u>length of</u> C/C_1 <u>is also defined, but may be</u> $< lg(C,C_1)$.

We define generally $lg(C,C_1)$ for any two lattices by putting

4.31 $$lg(C,C_1) = lg(C_o,C_1) - lg(C_o,C) \; ,$$

where C_o is a lattice in $\tilde{C}$ containing both C and C_1 : it does not depend on the choice of C_o , because

4.32 $$lg(C,C') + lg(C',C'') = lg(C,C'')$$

for any three lattices $C \supset C' \supset C''$ in $\tilde{C}$. Formula (4.28) results from (4.31).

5. <u>Unipotent formal groups</u>

5.1 In this section, $\tilde{C}$ <u>will denote a finite</u> $\tilde{E}$-<u>module</u> (see 4.20) over a perfect field K , and d <u>its dimension</u>, i.e. $\dim_K C/VC$, where C is any lattice in $\tilde{C}$.

5.2 A direct decomposition $\tilde{C} = \tilde{C}_1 \oplus \tilde{C}_2$ of $\tilde{C}$ qua $\tilde{E}$-module does not imply a corresponding direct decomposition of any lattice C . More precisely

5.3 $$C' = (C\cap\tilde{C}_1) \oplus (C\cap\tilde{C}_2)$$

is a direct decomposition of a lattice $C'\subset C$ qua E-module, but it may happen that $C' \neq C$ (see example 4.5).

5.4 We define a <u>semi-linear endomorphism</u> Λ of $\tilde{C}$ as an additive map, $\Lambda : \tilde{C} \to \tilde{C}$, together with an <u>automorphism</u> φ of K, which we extend to $E, E_+, \tilde{E}$: if $x = \Sigma_{m,n} V^m[a_{m,n}]F^n \in \tilde{E}$, then $\varphi(x) = \Sigma_{m,n} V^m[\varphi(a_{m,n})]F^n$. The relation between Λ and φ is

5.5 $$\Lambda(x\cdot\gamma) = \varphi(x)\ \Lambda(\gamma) \quad , \quad \text{for any } x \in \tilde{E}\ ,\ \gamma \in \tilde{C}\ .$$

If Λ, Λ' are so related to φ, φ' respectively, then $\Lambda\circ\Lambda'$ is a semi-linear endomorphism related to the automorphism $\varphi\circ\varphi'$ of K .

5.6 The reason for requiring that φ be an isomorphism is that we want Λ to map any submodule M of $\tilde{C}$ on a submodule $\Lambda.M$ (here, "submodules" are meant over W, E_+, E or $\tilde{E}$). The kernel of Λ is an $\tilde{E}$-submodule (even if φ is but an endomorphism of K).

5.7 <u>To any semi-linear endomorphism</u> Λ <u>of</u> $\tilde{C}$ <u>there corresponds a direct decomposition</u> $\tilde{C} = \tilde{C}_{nil} \oplus \tilde{C}_{bij}$, <u>defined by the condition that</u> Λ <u>is nilpotent on</u> $\tilde{C}_{nil}$ <u>and bijective on</u> $\tilde{C}_{bij}$.

5.8 This is Fitting's lemma, which applies generally to a semi-linear endomorphism Λ of a module M of finite (Jordan-Hölder) length, over any ring Ω whatsoever. We recall briefly its proof.

The increasing sequence of submodules $\operatorname{Ker}\Lambda \subset \operatorname{Ker}\Lambda^2 \subset \dots$ must have a stop, say at $\operatorname{Ker}\Lambda^\nu = \operatorname{Ker}\Lambda^{\nu+1} = \dots$, for M has finite length. Then, if $\gamma \in \operatorname{Ker}\Lambda^\nu \cap \operatorname{Im}\Lambda^\nu$, there is $\gamma' \in M$ such that $\gamma = \Lambda^\nu\gamma'$, $\Lambda^{2\nu}\gamma'$, therefore $\gamma = 0$ because $\operatorname{Ker}\Lambda^\nu = \operatorname{Ker}\Lambda^{2\nu}$. The submodule $\operatorname{Ker}\Lambda^\nu \oplus \operatorname{Im}\Lambda^\nu$ has the same length as M (because $\operatorname{Im}\Lambda^\nu \simeq M/\operatorname{Ker}\Lambda^\nu$), therefore is M . As $\operatorname{Ker}\Lambda \cap \operatorname{Im}\Lambda^\nu = 0$, Λ is bijective on $\operatorname{Im}(\Lambda^\nu)$.

5.9 For any $n \in \underline{\underline{P}}$, the n-dimensional formal group defined by the presentation

$$F\gamma_i = \gamma_{i+1} \quad (1\leq i\leq n-1) \ , \ F\gamma_n = 0 \ ,$$

is called the <u>formal group</u> of (classical, truncated) <u>Witt vectors of length</u> n , and denoted by $\hat{W}_n^+$. The corresponding group law is obtained by taking the addition formulas for Witt vectors and disregarding the components of index $\geq n$ (see III.2.15).

5.10 Let us assume that F is <u>nilpotent</u> on $\tilde{C}$ and, for any $\gamma \in \tilde{C}$, let $o_F(\gamma)$ denote the smallest integer n such that $F^n\gamma = 0$. By (1.6) any element of $E\gamma$ is written as

$$\Sigma_{n\in\underline{\underline{N}}, o\leq j<o_F(\gamma)} \ V^n[a_{n,j}]F^j.\gamma \qquad a_{n,j} \in K \ .$$

Let us choose a minimal set $(\gamma_1,\ldots,\gamma_r)$ of generators of the $\tilde{E}$-module $\tilde{C}$ which minimizes the sum

5.11 $$\Sigma_{1\leq i\leq r} \ o_F(\gamma_i) \ ,$$

and order them so that

5.12 $$o_F(\gamma_1) \geq o_F(\gamma_2) \geq \ldots \geq o_F(\gamma_r) \ .$$

Let C the E-submodule of $\tilde{C}$ generated by $(\gamma_1,\ldots,\gamma_r)$. Then C is a lattice in $\tilde{C}$ (just because the γ_i generate $\tilde{C}$) and we shall presently show that <u>the elements</u>

5.13 $$F^j.\gamma_i \ , \ 1\leq i\leq r \ , \ o\leq j<o_F(\gamma_i)$$

<u>are a</u> V-<u>basis of</u> C . This will prove that <u>the formal group</u> G <u>corresponding to</u> C <u>is a direct sum of</u> r <u>groups of Witt vectors, of respective lengths</u> $o_F(\gamma_1),\ldots, o_F(\gamma_2)$.

We have to prove that a non trivial relation

5.14 $$\Sigma_{n\in\underline{\underline{N}}, 1\leq i\leq r, o\leq j<o_i} \ V^n[a_{n,i,j}]F^j.\gamma_i = 0 \ , \ a_{n,i,j} \in K \ ,$$

implies a contradiction; we have written simply o_i instead of $o_F(\gamma_i)$.

By multiplying (5.14) by a suitable power of F , we may assume that the elements $F^j \cdot \gamma_i$ with non-vanishing coefficient in (5.14) lie in $\operatorname{Ker} F$, i.e. that $j = O_i - 1$. So (5.14) is replaced by

5.15 $$\Sigma_{n\in\underline{\underline{N}},1\leq i\leq r} V^n[a_{n,i}]F^{O_i-1}\cdot\gamma_i = 0 \quad , \quad a_{n,i} \in K \ .$$

Let α be the greatest index i in (5.15) for which $\Sigma\, V^n[a_{n,i}] \neq 0$, so that (5.15) may be rewritten as

5.16 $$\Sigma_{n\in\underline{\underline{N}},1\leq i\leq \alpha} V^n[a_{n,i}]F^{O_i-1}\cdot\gamma_i \ .$$

By eventually replacing the γ_i , where $i < \alpha$, by $V^\nu\gamma_i$, where $\nu \in \underline{\underline{N}}$ is large enough, we may assume that $a_{0,\alpha} \neq 0$ in (5.16): such a change does not alter the minimality of (5.11), because $O_F(V.\gamma) = O_F(\gamma)$, and allows us to factor out a power of V in (5.16), so that, after re-indexing, $a_{0,\alpha}$ becomes $\neq 0$. But then the coefficient of $F^{O_\alpha-1}\cdot\gamma_\alpha$ in (5.16) is invertible (see 2.13), and we have a relation of the form

5.17 $$F^{O_\alpha-1}\cdot\gamma_\alpha = \Sigma_{n\in\underline{\underline{N}},1\leq i<\alpha} V^n[b_{n,i}]F^{O_i-1}\cdot\gamma_i \ , \quad b_{n,i} \in K \ .$$

Let us put (see 5.12)

5.18 $$\gamma'_\alpha = \gamma_\alpha - \Sigma_{n\in\underline{\underline{N}},1\leq i<\alpha} V^n[b_{n,i}^{p^{1-O_\alpha}}]F^{O_i-O_\alpha}\cdot\gamma_i \ .$$

By (5.17), $F^{O_\alpha-1}(\gamma'_\alpha) = 0$, i.e. $O_F(\gamma'_\alpha) < O_F(\gamma_\alpha)$. As γ'_α and the γ_i $(1\leq i\leq r \ , \ i \neq \alpha)$ generate $\tilde{C}$, the sum (5.11) was not minimal: a contradiction.

5.19 Definition. <u>A formal group</u> G <u>is said to be unipotent iff the operator</u> F <u>is nilpotent on</u> $\mathbb{C}_p(G)$.

5.20 Proposition. <u>A unipotent formal group is isogenic to a direct sum of groups of Witt vectors</u> (see 5.9).

6. Spectral decomposition of semi-linear automorphisms of finite $\tilde{E}$-modules

6.1 Let $\tilde{C}$ denote again a finite $\tilde{E}$-module of dimension d over a perfect field K. We shall be interested in semi-linear endomorphisms and automorphisms of $\tilde{C}$, in the sense of (5.4), and we shall omit to repeat "semi-linear".

6.2 The order of an endomorphism Λ relatively to a lattice C in $\tilde{C}$, denoted by $\mathrm{ord}_C(\Lambda)$, is defined by

$$\mathrm{ord}_C(\Lambda) = \min_{\gamma\in C} \mathrm{ord}_C(\Lambda\cdot\gamma) \in \underline{\underline{Z}} \cup \{+\infty\} \qquad \text{(see 4.24)} .$$

Equivalently, $\mathrm{ord}_C(\Lambda)$ is (for $\Lambda \neq 0$), the greatest integer such that

6.3 $$\mathrm{ord}_C(\Lambda\cdot\gamma) \geqslant \mathrm{ord}_C(\Lambda) + \mathrm{ord}_C(\gamma) \quad , \quad \text{for any} \quad \gamma \in \tilde{C} .$$

We have

6.4 $$\mathrm{ord}_C(V^n\Lambda) = n + \mathrm{ord}_C(\Lambda) \quad , \quad \text{for any} \quad n \in \underline{\underline{Z}} ,$$

and

6.5 $$\mathrm{ord}_C(\Lambda\Lambda') \geqslant \mathrm{ord}_C(\Lambda) + \mathrm{ord}_C(\Lambda') \quad , \quad \text{for any two endomorphisms} \quad \Lambda,\Lambda' .$$

That implies

6.6 $$\mathrm{ord}_C(\Lambda^n) \geqslant n\, \mathrm{ord}_C(\Lambda) \quad , \quad \text{for any} \quad n \in \underline{\underline{P}} ,$$

6.7 For any lattice C in $\tilde{C}$, we shall denote by $\mathfrak{T}_C$ the natural map of C onto the K-vector space $\mathfrak{T}_C C = C/VC$; also for an endomorphism Λ such that $\mathrm{ord}_C(\Lambda) \geqslant 0$, i.e. $\Lambda C \subset \Lambda$, $\mathfrak{T}_C\Lambda$ will denote the semi-linear endomorphism of $\mathfrak{T}_C C$ induced by Λ :

6.8 $$\mathfrak{T}_C(\Lambda\cdot\gamma) = (\mathfrak{T}_C\Lambda)\cdot(\mathfrak{T}_C\gamma) \quad , \quad \gamma \in \tilde{C} ,$$

6.9 Lemma. Let $\Lambda \neq 0$ be an endomorphism of $\tilde{C}$ and C a lattice. Then either $n^{-1}\,\mathrm{ord}_C(\Lambda^n) = \mathrm{ord}_C(\Lambda)$ for any $n \in \underline{\underline{P}}$,

<u>or</u> $d^{-1}\mathrm{ord}_C(\Lambda^d) \geq \mathrm{ord}_C(\Lambda) + d^{-1}$.

Proof. By (6.4), we may assume that $\mathrm{ord}_C(\Lambda) = 0$, i.e. $\Lambda C \subset C$ and $\mathfrak{T}_C\Lambda \neq 0$. As $(\mathfrak{T}_C\Lambda)^n = \mathfrak{T}_C(\Lambda^n)$ for $n \in \underline{\underline{P}}$, either $\mathrm{ord}_C(\Lambda^n) = 0$ for any $n \in \underline{\underline{P}}$ or $\mathfrak{T}_C\Lambda$ is nilpotent. But as $\dim_K \mathfrak{T}_C C = d$, we have then $\mathfrak{T}_C(\Lambda^d) = 0$, i.e. $\mathrm{ord}_C(\Lambda^d) \geq 1$.

6.10 Among endomorphisms of $\tilde{C}$, <u>automorphisms</u> are characterized equivalently as being injective, surjective, or bijective, for $\tilde{C}$ has <u>finite length</u>. An automorphism Λ of $\tilde{C}$ maps any lattice C onto a lattice ΛC ; moreover

6.11 $$\mathrm{lg}(\Lambda C, \Lambda C') = \mathrm{lg}(C, C') \qquad \text{(see 4.23)}$$

for any two lattices in $\tilde{C}$: it suffices to prove (6.11) when $C' \subset C$, and then Λ induces an isomorphism of the finite E_+-module C/C' onto $\Lambda C/\Lambda C'$.

By (4.28) and (6.11), the number

6.12 $$\mathrm{lg}(C, \Lambda C) = \mathrm{lg}(\Lambda) \in \underline{\underline{Z}}$$

<u>does not depend on the lattice</u> C : we call it the <u>length</u> of Λ . We have

6.13 $$\mathrm{lg}(V) = d \qquad \text{(by definition)} ,$$

and, for any two automorphisms of $\tilde{C}$,

6.14 $$\mathrm{lg}(\Lambda\Lambda') = \mathrm{lg}(\Lambda) + \mathrm{lg}(\Lambda') ,$$

because

$$\mathrm{lg}(\Lambda\Lambda') = \mathrm{lg}(C, \Lambda C) + \mathrm{lg}(\Lambda C, \Lambda\Lambda' C) = \mathrm{lg}(\Lambda) + \mathrm{lg}(\Lambda') \quad \text{(see 6.11)}.$$

6.15 <u>Lemma</u>. <u>For any automorphism</u> Λ <u>of</u> $\tilde{C}$, <u>any lattice</u> C <u>in</u> $\tilde{C}$ <u>and any</u> $n \in \underline{\underline{P}}$, <u>we have</u>

6.16 $$n^{-1}\mathrm{ord}_C(\Lambda^n) \leq d^{-1}\mathrm{lg}(\Lambda) .$$

Proof. By (6.14), we have $\mathrm{lg}(\Lambda^n) = n\,\mathrm{lg}(\Lambda)$ for any $n \in \underline{\underline{P}}$, so that it suffices to prove (6.16) for $n = 1$. But, by (6.4), (6.13), (6.14),

$$d^{-1}\lg(\Lambda) - \mathrm{ord}_C(\Lambda) = d^{-1}\lg(V^n\Lambda) - \mathrm{ord}_C(V^n\Lambda)$$

for any $n \in \underline{\underline{Z}}$, so that we may assume that $\mathrm{ord}_C(\Lambda) = 0$; then $\Lambda C \subset C$, and, obviously, $d^{-1}\lg(\Lambda) \geq 0 = \mathrm{ord}_C(\Lambda)$.

6.17 The spectral order of an endomorphism Λ of $\tilde{C}$, denoted by $\mathrm{ord}_{sp}(\Lambda)$, is defined by analogy with the spectral norm in Banach algebras: see e.g. [12], p. 75. Let C be any lattice in $\tilde{C}$. Then either Λ is nilpotent (and we put $\mathrm{ord}_{sp}(\Lambda) = +\infty$) , or there is an integer $\nu \in \underline{\underline{P}}$ such that

6.18 $$\nu^{-1}\mathrm{ord}_C(\Lambda^\nu) = (n\nu)^{-1}\,\mathrm{ord}_C(\Lambda^{n\nu}) \text{ , for any } n \in \underline{\underline{P}} .$$

Then the rational number

6.19 $$\mathrm{ord}_{sp}(\Lambda) = \nu^{-1}\mathrm{ord}_C(\Lambda^\nu)$$

does not depend on the lattice C ; it is defined equivalently by

6.20 $$\mathrm{ord}_{sp}(\Lambda) = \sup_{n\in\underline{\underline{P}}}\ n^{-1}\mathrm{ord}_C(\Lambda^n)$$

or by

6.21 $$\mathrm{ord}_{sp}(\Lambda) = \lim_{n\to\infty}\ n^{-1}\mathrm{ord}_C(\Lambda^n) .$$

We have

6.22 $$\mathrm{ord}_{sp}(\Lambda^n) = n\ \mathrm{ord}_{sp}(\Lambda) \quad \text{for any } n \in \underline{\underline{N}} ,$$

and

6.23 $\mathrm{ord}_{sp}(\Lambda) \geq 0$ iff there is a lattice C' in $\tilde{C}$ such that $\Lambda C' \subset C'$.

Proof. By (5.7), we may assume that Λ is an automorphism. Then lemma (6.15) shows that the sequence $n^{-1}\mathrm{ord}_C(\Lambda^n)$ is bounded from above. We take (6.20) as a definition of $\mathrm{ord}_{sp}(\Lambda)$; then it follows from (4.26) that the upper bound of $n^{-1}\mathrm{ord}_C(\Lambda^n)$

does not depend on the lattice C . If we take $\nu \in \underline{\underline{P}}$ such that $\mathrm{ord}_{sp}(\Lambda) - \nu^{-1}\mathrm{ord}_c(\Lambda^\nu) < d^{-1}$, then lemma (6.9) implies (6.18) and (6.19).

To show that (6.20) implies (6.21), we divide n by ν , $n = q\nu + r$, $0 \leq r < \nu$, and we have

$$\mathrm{ord}_c(\Lambda^n) \geq \mathrm{ord}_c(\Lambda^{q\nu}) + \mathrm{ord}_c(\Lambda^r) ,$$

$$\geq q\nu\, \mathrm{ord}_{sp}(\Lambda) - \nu|\mathrm{ord}_c(\Lambda)| ,$$

whence

$$n^{-1}\mathrm{ord}_c(\Lambda^n) \geq \mathrm{ord}_{sp}(\Lambda) - n^{-1}\nu(|\mathrm{ord}_{sp}(\Lambda)|+|\mathrm{ord}_c(\Lambda)|).$$

The relation (6.22) follows from (6.6) and (6.20). Finally, let us put

6.24 $$C' = \Sigma_{n\in \underline{\underline{N}}} \Lambda^n C .$$

If $\mathrm{ord}_{sp}(\Lambda) \geq 0$ and $\nu^{-1}\mathrm{ord}_c(\Lambda^\nu) = \mathrm{ord}_{sp}\Lambda$, then $\Lambda^\nu C \subset C$, so that (6.24) becomes a finite sum:

6.25 $$C' = \Sigma_{0\leq n<\nu} \Lambda^n C ,$$

the E-submodule C' contains C and is finitely generated, therefore it is a lattice verifying $\Lambda C' \subset C'$. Conversely, that implies $\mathrm{ord}_{c'}(\Lambda^n) \geq 0$ for every $n \in \underline{\underline{P}}$, so that $\mathrm{ord}_{sp}(\Lambda) \geq 0$.

6.26 Lemma. Let Λ be an endomorphism of $\tilde{C}$ and C a lattice such that $\mathrm{ord}_{sp}(\Lambda) = \mathrm{ord}_c(\Lambda) = 0$. Moreover, let us assume that the nilpotent component of $\mathfrak{T}_c\Lambda$ is 0 (see 5.8 and 6.7). So we have a direct decomposition,

$$\mathfrak{T}C = T_0 \oplus T_1 , \quad T_0 \neq 0 ,$$

such that $\mathfrak{T}_c\Lambda$ is bijective on T_0 and zero on T_1 .

Then there is a direct decomposition of the E-module C ,

$$C = C_0 \oplus C_1 ,$$

such that

6.27 $$\mathfrak{T}_c C_o = T_o \ , \quad \mathfrak{T}_c C_1 = T_1 \ , \quad \Lambda C_o \subset C_o \ , \quad \Lambda C_1 \subset C_1 \ ;$$

6.28 <u>for any</u> $\gamma \in C$, $\gamma \in C_1$ <u>iff</u> $\mathrm{ord}_c(\Lambda^n \gamma) \geq n$ <u>for any</u> $n \in \underline{\underline{P}}$;

6.29 $$C_o = \cap_{n \in \underline{\underline{P}}} \Lambda^n C \ .$$

Proof. For any $n \in \underline{\underline{P}}$, let us define $C_{1,n} \subset C$ by

6.30 $$\gamma \in C_{1,n} \text{ iff } \mathrm{ord}_c(\Lambda^i \gamma) \geq i \text{ for } 1 \leq i \leq n \ .$$

Clearly, $C_{1,n}$ is a closed E-submodule of C , and C_1

$C_1 = \cap_{n \in \underline{P}} C_{1,n}$.

Assume that

6.31 $$\gamma \in C_{1,n} \ , \quad \gamma \notin C_{1,n+1} \ .$$

Then, as

6.32 $$\mathrm{ord}_c(\Lambda^{n+1}\gamma) \geq \mathrm{ord}_c(\Lambda^n \gamma) \geq n \ ,$$

we must have

6.33 $$n = \mathrm{ord}_c(\Lambda^n \gamma) = \mathrm{ord}_c(\Lambda^{n+1}\gamma)$$

Putting

$$\delta = V^{-n}\Lambda^n \gamma \in C \ ,$$

we have $0 = \mathrm{ord}_c(\delta) = \mathrm{ord}_c(\Lambda\delta)$, or equivalently

$$\mathfrak{T}_c \delta = t_o + t_1 \ , \quad t_o \in T_o \ , \ t_1 \in T_1 \ , \ t_o \neq 0 \ .$$

As $\mathfrak{T}_c \Lambda$ is bijective on T_o , there is a $\delta' \in C$ such that

$$\mathfrak{T}_c(\Lambda^n \delta') = t_o \ .$$

If we put

6.34 $$\gamma' = \gamma - V^n \delta' \ ,$$

we have $\gamma' \in C_{1,n}$ and

$$\mathfrak{T}_c(V^{-n}\Lambda^n \gamma') = t_1 \ ,$$

which implies $\mathfrak{T}_c(V^{-n}\Lambda^{n+1}\gamma') = 0$, i.e. $\gamma' \in C_{1,n+1}$.

Therefore, by successive approximations, we obtain the existence of a $\gamma \in C_1$ with $\mathfrak{T}_c \gamma_1 = v$, for any given $v \in T_1$.

As $\mathfrak{T}_c C_1 \subset T_1$, we have $\mathfrak{T}_c C_1 = T_1$.

Besides, if $\gamma \in C$ satisfies (6.31), we have $\mathrm{ord}_c(\Lambda^i \gamma) = n$ for any $i \geq n$. Therefore $V^m \gamma \notin C_1$ for any $m \in \underline{\underline{N}}$, i.e. C_1 is V-divisible in C .

So C/C_1 is a reduced E-module (see 3.9), and we denote by $\bar{\Lambda}$ the action of Λ on C/C_1 (obviously $\Lambda C_1 \subset C_1$). We may identify $\mathfrak{T}(C/C_1)$ with T_o , which shows that $\bar{\Lambda}$ is an <u>automorphism</u> of the E-module C/C_1. Therefore, any $\bar{\gamma} \in C/C_1$ has a representative $\gamma_n \in \Lambda^n C$, and any two such representatives, γ_n, γ_n' , verify $\mathrm{ord}_c(\gamma_n - \gamma_n') \geq n$. Putting $\gamma = \lim_{n\to\infty} \gamma_n$, we obtain the <u>unique representative of</u> γ <u>in</u> $\bigcap_{n \in \underline{\underline{N}}} \Lambda^n C$. Those distinguished representatives form an E-submodule C_o of C , such that $\mathfrak{T}_c C_o = T_o$, $\Lambda C_o \subset C_o$, $C_o \cap C_1 = 0$. The direct sum $C_o \oplus C_1$ is C, because $\mathfrak{T}C = \mathfrak{T}_c C_o \oplus \mathfrak{T}_c C_1$; $\Lambda^n C \subset C_o \oplus V^n C_1$ for any $n \in \underline{\underline{N}}$, so that $C_o = \bigcap_{n \in \underline{\underline{N}}} \Lambda^n C$.

6.35 <u>Proposition</u>. <u>For an automorphism</u> Λ <u>of</u> $\tilde{C}$, <u>the following three properties are equivalent</u>:

6.36
$$\mathrm{ord}_{sp}\Lambda = d^{-1}\, \lg(\Lambda) .$$

6.37 $\lim_{n\to\infty} n^{-1} \mathrm{ord}_c(\Lambda^n \gamma) = \mathrm{ord}_{sp}(\Lambda)$, <u>for any lattice</u> C <u>in</u> $\tilde{C}$ <u>and any</u> $\gamma \in \tilde{C}$, $\gamma \neq 0$;

6.38 <u>there is a lattice</u> C <u>in</u> $\tilde{C}$ <u>such that</u>

$$\mathrm{ord}_c(\Lambda^n \gamma) = n\, \mathrm{ord}_{sp}(\Lambda) + \mathrm{ord}_c(\gamma)$$

<u>for any</u> $\gamma \in \tilde{C}$, $\gamma \neq 0$ <u>and any</u> $n \in \underline{\underline{N}}$ <u>such that</u> $n\, \mathrm{ord}_{sp}(\Lambda) \in \underline{\underline{Z}}$.

Proof. Put $\mathrm{ord}_{sp}(\Lambda) = \mu/\nu$, with $\mu \in \underline{\underline{Z}}$, $\nu \in \underline{\underline{P}}$, $\gcd(\mu,\nu) = 1$, and

$$\Lambda_o = V^{-\mu}\Lambda^{\nu} .$$

Then we have

$$\mathrm{ord}_{sp}(\Lambda_o) = \nu\, \mathrm{ord}_{sp}(\Lambda) - \mu = 0 \quad \text{(see 6.4, 6.22)},$$

and

$$d^{-1}\lg(\Lambda_o) = \nu\, d^{-1}\lg(\Lambda) - \mu \qquad \text{(see 6.13, 6.14)} .$$

Therefore, condition (6.36) is equivalent to

$$\lg(\Lambda_o) = 0 .$$

By (6.23), there is a lattice C in $\tilde{C}$ such that $\Lambda_o C \subset C$. Then $\lg(\Lambda_o) = 0$ iff $\Lambda_o C = C$, or iff $\mathfrak{T}_c \Lambda_o$ is an automorphism of $\mathfrak{C}$. So (6.36) and (6.38) are equivalent, and imply (6.37) by (4.26).

If (6.36) and (6.38) are false, then we may apply lemma (6.26) to some power $\Lambda_1 = \Lambda_o^{\alpha}$ of Λ_o, and there are elements $\gamma \in C$ such that $\gamma \neq 0$ $\mathrm{ord}_c(\Lambda_1^n \gamma) \geq n$ for any $n \in \underline{\underline{P}}$, so that (6.37) does not hold.

6.39 <u>Definition</u>. <u>We say that an automorphism</u> Λ <u>of</u> $\tilde{C}$ <u>is isoclinal iff it verifies the equivalent conditions of proposition</u> (6.35).

6.40 <u>Theorem</u>. <u>For any automorphism</u> Λ <u>of</u> $\tilde{C}$, <u>there is a unique direct decomposition</u> $\tilde{C} = \oplus_{1 \leq i \leq n} \tilde{C}_i$ <u>with the following properties</u> : $\Lambda\tilde{C}_i \subset \tilde{C}_i$ <u>and the restriction</u> Λ_i <u>of</u> Λ <u>to</u> $\tilde{C}_i$ <u>is isoclinal for</u> $1 \leq i \leq n$; $\mathrm{ord}_{sp}\Lambda_i \neq \mathrm{ord}_{sp}\Lambda_j$ for $i \neq j$.

Proof by induction on d : we argue as in the proof of proposition (6.35) and we apply lemma (6.26) to Λ_1, in order to split off the Λ-isoclinal component of $\tilde{C}$ on which Λ has the least spectral order.

7. <u>Formal groups of finite height</u>

7.1 <u>Proposition</u>. <u>Let</u> G <u>be a formal group of finite dimension</u> d <u>over a perfect field</u> K . <u>Then the following properties are equivalent</u>:

7.2 <u>the multiplication by</u> p <u>is an isogeny of</u> G <u>into itself</u> (see 4.21);

7.3 <u>if</u> $\gamma \in \mathbb{C}_p(G)$ <u>and</u> $F\cdot\gamma = 0$, <u>then</u> $\gamma = 0$;

7.4 $\mathbb{C}_p(G)$ <u>is a free</u> W-<u>module of finite rank</u>.

Proof. Let $C = \mathbb{C}_p(G)$ and $\tilde{C}$ be the V-divided module of C. Then (7.2) holds iff F is an automorphism of $\tilde{C}$ or, by (6.10), iff P is injective (on $\tilde{C}$ or equivalently on C): so (7.2) and (7.3) are equivalent.

As $p = VF = FV$, we have $\mathrm{Ker}_C F = \mathrm{Ker}_C p$, so that, if (7.3) does not hold, C contains a submodule isomorphic to $\mathbb{C}_p(K^+)$ and is neither p-torsion-free nor finitely generated over the noetherian ring W.

Conversely, if (7.3) hold, F has a length in $\tilde{C}$:

7.5 $$d' = \lg(F) = \lg(C,FC) = \dim_K C/FC \qquad \text{(see 6.12)},$$

and

7.6 $$\dim_K(C/pC) = \lg(V) + \lg(F) = d + d' = h \qquad \text{(see 6.14)}.$$

Let $(\gamma_j)_{1\le j\le h}$ be representatives of a basis of C/pC over K. As C is p-torsion-free, there are, for any $\gamma \in C$, unique coefficients $a_{n,j} \in K$, $n \in N$, $1 \le j \le d$, such that

7.7 $$\gamma - \Sigma_{o\le i<n,1\le j\le h}\, p^i[a_{i,j}]\cdot\gamma_j \in p^n C, \qquad \text{for any } n \in \underline{\underline{P}}$$

As C is complete for its p-adic topology (see 2.5) and as any Witt vector $\xi \in W$ has a unique representation in the form $\Sigma_{i\in \underline{\underline{N}}}\, p^i[a_i]$, $a_i \in K$ (see 3.13), we have proved that C <u>is a free</u> W-<u>module of rank</u> h, <u>with basis</u> $(\gamma_j)_{1\le j\le h}$ <u>over</u> W.

7.8 <u>Remark</u>. A basis of C over W is not a V-basis of C, unless $d' = 0$. By applying proposition (3.17) to the submodule pC of C, we see that there is a V-basis $(\gamma_i)_{1\le i\le d}$ of C

and numbers $n_i \in \underline{\underline{P}}$, such that the curves $V^{\alpha}\cdot\gamma_i$ with $0 \leq \alpha < n_i$, $1 \leq i \leq d$, are a basis of C over $W(\Sigma_{1\leq i\leq d} n_i = h)$.

7.9 Definitions. A formal group G has finite height iff it verifies the conditions of (7.1). Then its height h is the rank of the free module $\mathbb{C}_p(G)$ over W ,

7.10 $$h = d + d' ,$$

where d is the dimension of G and $d' = \dim_K \mathbb{C}_p(G)F\mathbb{C}_p(G)$ is called the codimension of G .

An isoclinal formal group is a formal group G of finite height such that the semi-linear isomorphism F of the V-divided module of $\mathbb{C}_p(G)$ is isoclinal in the sense of (6.3.9), i.e.

7.11 $$\lim_{n\to\infty} n^{-1}\mathrm{ord}(F^n\gamma) = d^{-1}d' \quad \text{for any} \quad \gamma \in \mathbb{C}_p(G),\ \gamma \neq 0 .$$

Let G be any finite-dimensional formal group over the perfect field K . Then, by (5.7) and (5.20), G is isogenic to a direct sum of groups of Witt vectors and of a group of finite height. By (6.40), this latter is isogenic to a direct sum of isoclinal groups.

7.12 Theorem. A finite dimensional formal group over a perfect field is isogenic to a direct sum of additive groups of Witt vectors and of isoclinal groups.

To prove the existence of an isoclinal formal group with prescribed values of $d \in \underline{\underline{P}}$ and $d' \in \underline{\underline{N}}$, it suffices to take the group defined by the basic set of curves $(\gamma_i)_{1\leq i\leq d}$ and the relations

7.13 $$\begin{cases} F\cdot\gamma_i = \gamma_{i+1} , & 1 \leq i < d , \\ F\cdot\gamma_d = V^{d'}\gamma_1 . \end{cases}$$

Then $F^d \cdot \gamma_i = V^{d'} \cdot \gamma_i$ for $1 \leq i \leq d$, and $\mathrm{ord}_{sp}(F) = d^{-1}d'$.

7.14 From now on (until the end of the present chapter), we shall consider only isoclinal formal groups with a fixed value

$\varrho = \mu^{-1}\nu$ of $\mathrm{ord}_{sp}(F)$, where $\gcd(\mu,\nu) = 1$ ($\mu = 1$ if $\nu=0$).

We put

7.15 $$\Lambda = V^{-\nu}F^{\mu} \in \tilde{E} ,$$

so that the finite $\tilde{E}$-modules $\tilde{C}$ under investigation are characterized by the condition

7.16 $$\lim_{n\to\infty} n^{-1}\mathrm{ord}_c(\Lambda^n \cdot \gamma) = 0 \quad \text{for any} \quad \gamma \in \tilde{C} , \ \gamma \neq 0 .$$

7.17 Condition (7.16) is shared by any $\tilde{E}$-submodule or factor module of $\tilde{C}$. As the dimension d of $\tilde{C}$ must be an integral multiple of μ (because $d^{-1}d' = \mu^{-1}\nu$ and $\gcd(\mu,\nu) = 1$), $\tilde{C}$ is simple if its dimension is μ .

7.18 Proposition. An isoclinal formal group G of dimension d and codimension d', with $\gcd(d,d') = 1$, is isogenically simple, i.e. any embedded subgroup of G is 0 or G .

It is convenient to replace the spectral order, which depends on the dimension, by the p-adic valuation, denoted by ord_p , defined as the spectral order divided by the height. So we put

7.19 $$\mathrm{ord}_p(p) = 1 , \quad \mathrm{ord}_p(V) = (\mu+\nu)^{-1}\mu , \quad \mathrm{ord}_p(F) = (\mu+\nu)^{-1}\nu .$$

Let $\alpha,\beta \in \underline{\underline{N}}$ satisfy

7.20 $$\alpha\nu - \beta\mu = 1$$

(we could define them completely by adding the condition $0 \leq \alpha < \mu$),

and let us put

7.21 $$\theta = V^{-\beta}F^{\alpha} \in \tilde{E} .$$

Then we have

7.22 $$\mathrm{ord}_p(\theta) = (\mu+\nu)^{-1} \quad , \quad \mathrm{ord}_p(\Lambda) = 0 \ ,$$

7.23 $$V = \theta^{\mu}\Lambda^{-\alpha} \quad , \quad F = \theta^{\nu}\Lambda^{-\beta} \quad , \quad p = \theta^{\mu+\nu}\Lambda^{-\alpha-\beta} \quad \text{in} \quad \tilde{E} \ .$$

7.24 Proposition. There are lattices C in $\tilde{C}$ such that

7.25 $$\Lambda C = C \quad , \quad \theta C \subset C \ .$$

That follows from (6.23), because Λ and θ commute (see 6.25). An equivalent way to state proposition (7.24) is to say that there are lattices which are E_1-modules, where

7.26 $$E_1 = E[\Lambda, \Lambda^{-1}, \theta)$$

is obtained by adjoining Λ, Λ^{-1} and θ to E .

7.27 Definition. Let $G_{\mu,\nu}$ be the formal group of dimension μ such that $\mathfrak{C}_p(G_{\mu,\nu})$ is generated, qua E_1-module, by a curve γ verifying $\Lambda\cdot\gamma = \gamma$.

We put

7.28 $$\gamma_i = \theta^{i-1}\gamma \quad \text{for} \quad i \in \underline{\underline{P}} \ .$$

Then we have, by (7.23)

$$\Lambda\cdot\gamma_i = \gamma_i \quad , \quad V\cdot\gamma_i = \gamma_{i+\mu} \quad , \quad F\cdot\gamma_i = \gamma_{i+\nu} \quad , \quad p\gamma_i = \gamma_{i+\mu+\nu},$$

for any $i \in \underline{\underline{P}}$. The curves $(\gamma_i)_{1\leq i\leq\mu}$ are a V-basis in $\mathfrak{C}_p(G_{\mu,\nu})$, and the corresponding presentation is

7.29 $$F\cdot\gamma_i = V^{\alpha}\gamma_{i+\nu-\alpha\mu} \quad ,$$

where $\alpha \in \underline{\underline{N}}$ is defined by $1 \leq i + \nu - \alpha\mu \leq \mu$.

We may consider that $G_{\mu,\nu}$ is defined as a p-typical formal group over $\underline{\underline{Z}}$ by (7.29) and then reduced modulo p (see VII.1), to become a formal group over $\underline{\underline{F}}_p$ or any ring of characteristic p .

7.30 We put $C_{\mu,\nu} = \mathfrak{C}_p(G_{\mu,\nu})$, and the V-divided $\tilde{E}$-module $\tilde{C}_{\mu,\nu}$

of $C_{\mu,\nu}$ is simple, by (7.18), so that its ring of endomorphisms is a division algebra, by Schur's lemma.

The endomorphisms of $C_{\mu,\nu}$ (resp. $\tilde{C}_{\mu,\nu}$) are in one-to-one correspondence with the curves $\gamma' \in C_{\mu,\nu}$ (resp. $\tilde{C}_{\mu,\nu}$) such that $\Lambda\gamma' = \gamma'$ (the corresponding endomorphism mapping $\gamma = \gamma_1$ on γ'). We write a general curve γ' as

7.31 $$\gamma' = \Sigma_{i,n} V^n[a_{n,i}]\cdot\gamma_i \quad , \quad a_{n,i} \in K \; ,$$

where $1 \leq i \leq \mu$, $n \in \underline{\underline{N}}$ iff $\gamma' \in C_{\mu,\nu}$, $n \in \underline{\underline{Z}}$, $n \geq n_o$ if $\gamma' \in \tilde{C}_{\mu,\nu}$.

Let us put

7.32 $$q = p^{\mu+\nu} \; ,$$

and denote by $\underline{\underline{F}}_q$ the finite field with q elements (i.e. the splitting field of the polynomial $X^q - X$). Then, as

$$\Lambda[a] = [a^q]\Lambda \; , \quad \text{for any } a \in K \; ,$$

by (7.15), we have

7.33 $$\Lambda\cdot\gamma' = \gamma' \quad \text{iff} \quad a_{n,i} \in K \cap \underline{\underline{F}}_q \quad \text{for any } n,i \; .$$

Therefore, all possible endomorphisms of $C_{\mu,\nu}$ (resp. $\tilde{C}_{\mu,\nu}$) are obtained when

7.34 $$K = \underline{\underline{F}}_q \; ,$$

which we assume from now on. Then the lattices we are considering become lattices as in [15], chp. 2, $C_{\mu,\nu}$ becomes compact and $\tilde{C}_{\mu,\nu}$ locally compact.

7.35 Let us denote by Ω (resp. R) the division algebra of endomorphisms of $\tilde{C}_{\mu,\nu}$ over $\underline{\underline{F}}_q$ (resp. the ring of endomorphisms of $C_{\mu,\nu}$). We let Ω act on the right, in order to define an embedding of $W(\underline{\underline{F}}_q)$ in R by putting

7.36 $$\gamma_1\cdot\xi = \xi\cdot\gamma_1 \quad , \quad \text{for any } \xi \in W(\underline{\underline{F}}_q) \; .$$

Note that the embedding depends on the choice of γ_1 , i.e. we do <u>not</u> have $\gamma\cdot\xi = \xi\cdot\gamma$ for any $\gamma \in C_{\mu,\nu}$.

We define $\omega_V, \omega_F, \omega_\theta \in R$ by

7.37 $$\gamma_1\cdot\omega_V = V\cdot\gamma_1 \quad , \quad \gamma_1\cdot\omega_F = F\cdot\gamma_1 \quad , \quad \gamma_1\cdot\omega_\theta = \theta\cdot\gamma_1 \ .$$

Then we have the <u>commutation rules</u>

7.38 $$\xi\omega_V = \omega_V\xi^F \quad , \quad \omega_F\xi = \xi^F\omega_F \quad ,$$

<u>and the relations</u>

7.39 $$\omega_V = \omega_\theta^\mu \quad , \quad \omega_F = \omega_\theta^\nu \quad , \quad p = \omega_\theta^{\mu+\nu}$$

in R . <u>Any element of</u> R <u>has a unique expansion in the form</u>

7.40 $$\Sigma_{n\in\underline{\underline{N}}}\ \omega_\theta^n\,[a_n] \quad , \quad a_n \in \underline{\underline{F}}_q \ ,$$

because it maps γ_1 on the curve $\Sigma_{n\in\underline{\underline{N}}}\ \theta^n[a_n]\cdot\gamma_1$, a general expression equivalent to (7.31).

Similarly, any element of Ω has a unique expansion as

7.41 $$\Sigma_{n\geqslant -n_o, n\in\underline{\underline{Z}}}\ \omega_\theta^n[a_n] \quad , \quad a_n \in F_q \ .$$

It follows from the commutation rules (7.38) that the center of R is $W(\underline{\underline{F}}_p) = \underline{\underline{Z}}_p$ (the ring of p-adic integers), and that the center of Ω is $\underline{\underline{Q}}_p$ (the field of p-adic numbers). We have taken care to embed $W(\underline{\underline{F}}_q)$ in R (generally speaking, the embedding is only defined up to inner automorphism of Ω). The field of quotients of $W(\underline{\underline{F}}_q)$ is the <u>unramified extension of degree</u> $(\mu+\nu)$ <u>of</u> $\underline{\underline{Q}}_p$, <u>and</u> Ω <u>has rank</u> $(\mu+\nu)^2$ <u>over</u> $\underline{\underline{Q}}_p$.

All that is proved by simple computations, without reference to the general theory of division algebras. This latter says that a central finite-dimensional division algebra over $\underline{\underline{Q}}_p$ is characterized, up to inner automorphisms, by its <u>invariant</u>, which is a rational number modulo 1 , defined (with the present notations), as the p-adic valuation of an element $a \in \Omega$ such that $a^{-1}\xi a = \xi^F$ for any $\xi \in W(\mathbb{F}_q)$: see, e.g., [2], p. 125. By (7.38) and (7.39), we may take $a = \omega_F$, so that the invariant of Ω is $(\mu+\nu)^{-1}\nu$, mod. 1. We summarize our results as follows.

7.42 Proposition. Let Ω be the central division algebra of rank $(\mu+\nu)^2$ over $\mathbb{Q}_p$ and invariant $(\mu+\nu)^{-1}\nu$ mod. 1. Let R be the ring of integers of Ω (equivalently: its maximal compact subring, or the set of its elements with p-adic valuation ≥ 0). Then the ring of endomorphisms of the formal group $G_{\mu,\nu}$ (see 7.27) may be identified with R .

8. Isoclinal formal groups over an algebraically closed field

8.1 Theorem. Over an algebraically closed field K , all isoclinal formal groups where F has a given p-adic valuation $(\mu+\nu)^{-1}\nu$ (see 7.19) are classified, up to isogeny, by their dimension (an integral multiple of μ). In other words, they are isogenic to a direct sum of copies of the formal group $G_{\mu,\nu}$ (see 7.27).

8.2 By proposition (7.24), we may choose in any isogeny class a formal group G such that $C = \mathfrak{C}_p(G)$ verifies

8.3 $$\Lambda C = C \quad , \qquad \theta C \subset C \ ,$$

where as before (7.15), (7.21)

8.4 $$\Lambda = V^{-\nu}F^{\mu} \in \tilde{\mathbb{E}} \quad , \ \theta = V^{-\beta}F^{\alpha} \in \tilde{\mathbb{E}} \quad , \quad \alpha\nu-\beta\mu = 1 \ .$$

The proof of theorem (8.1) will consist in showing that there are enough curves in C invariant by Λ .

8.5 Lemma. Let Γ be a finite-dimensional vector space over a separably closed field K of characteristic p , let $h \in \mathbb{P}$

$q = p^h$. Let f be a semi-linear additive endomorphism of T, such that

$$f(av) = a^q f(v) \quad , \quad \text{for any} \quad a \in K, \ v \in T,$$

and $\operatorname{Ker} f = 0$. Then the set T_o of vectors $v \in T$, such that $f(v) = v$, is a vector space over $\underset{=}{F}_q$, and any basis of T_o over $\underset{=}{F}_q$ is also a basis of T over K (i.e. $T = K \underset{\underset{=}{F}_q}{\otimes} T_o$).

Proof. Choose any $v_o \neq 0$ in T and let r be the smallest integer such that the vectors

8.6
$$v_o, \ v_1 = f(v_o), \ldots, v_r = f(v_{r-1})$$

are linearly dependent over K. So there is a relation

8.7
$$\Sigma_{o \leqslant i \leqslant r} \ \lambda_i v_i = 0 \quad , \quad \lambda_i \in K,$$

with $\lambda_r \neq 0$, and also $\lambda_o \neq 0$ (because f is injective). Put

8.8
$$v = \Sigma_{o \leqslant i \leqslant r-1} \ x_i v_i \quad , \quad x_i \in K.$$

Then, by (8.6),

8.9
$$v - f(v) = \Sigma_{o \leqslant i \leqslant r} \ (x_i - x_{i-1}^q) v_i \quad , \quad x_{-1} = x_r = 0.$$

Therefore, we have $f(v) = v$ iff

8.10
$$x_i - x_{i-1}^q = \lambda_i y \quad \text{for} \quad 0 \leqslant i \leqslant r \quad , \quad x_{-1} = x_r = 0,$$

where y is some element in K. Equations (8.10) are equivalent to

8.11
$$x_i = \Sigma_{o \leqslant n \leqslant i} \ \lambda_n^{q^{i-n}} y^{q^{i-n}} \quad , \quad 0 \leqslant i \leqslant r-1$$

and

8.12
$$\Sigma_{o \leqslant n \leqslant r} \ \lambda_n^{q^{r-n}} y^{q^{r-n}} = 0.$$

So we have a separable equation of degree q^r for y; its roots lie in K and we have found non-zero f-invariant vectors in T (as a matter of fact, the lemma is proved if T is generated by v_o qua (K,f)-module). Let us assume, by induction on the dimension δ of T, that there is a basis $(e_1, e'_2, \ldots, e'_\delta)$ of T over K, where $f(e_1) = e_1$ and

8.13 $$f(e'_i) = e'_i + a_i e_1 \quad , \quad a_i \in K \ , \ 2 \leqslant i \leqslant \delta \ .$$

Then we put

8.14 $$e_i = e'_i + x_i e_1 \quad , \quad x_i \in K \quad , \quad 2 \leqslant i \leqslant \delta \ ,$$

and we have

8.15 $$f(e_i) - e_i = (a_i + x_i^q - x_i) e_1 \quad ,$$

so that we obtain an f-invariant basis $(e_1, e_2, \ldots, e_\delta)$ in T by solving in K the separable equations $a_i + x_i^q - x_i = 0$, and (8.5) is proved.

8.16 Now we take a formal group G with $C = \mathfrak{C}_p(G)$ verifying (8.3), and we put $T = C/\theta C$, f being the semi-linear automorphism of T induced by Λ. Let $\gamma \mapsto \bar{\gamma}$ be the natural map $C \to T$. By lemma (8.5), there are curves $\gamma_{1,o}, \ldots, \gamma_{\delta,o} \in C$ such that $f(\bar{\gamma}_{i,o}) = \bar{\gamma}_{i,o}$ and that the $\gamma_{i,o}$ are a basis of T over K $(1 \leqslant i \leqslant \delta)$. We construct the curves γ_i, invariant by Λ, by successive approximations. Let us assume that we have the curves $(\gamma_{i,n-1})$, $1 \leqslant i \leqslant \delta$, such that

$$\Lambda \cdot \gamma_{i,n-1} \equiv \gamma_{i,n-1} \quad \text{mod. } \theta^n C \ ,$$

say

8.17 $$\Lambda \cdot \gamma_{i,n-1} = \gamma_{i,n-1} + \theta^n \varepsilon_i \quad , \quad \varepsilon_i \in C \ .$$

We put

8.18 $$\gamma_{i,n} = \gamma_{i,n-1} + \theta^n \cdot \eta_i \quad , \quad \eta_i \in C \ ,$$

and we have

$$\Lambda \cdot \gamma_{i,n} = \gamma_{i,n} + \theta^n (\varepsilon_i + \Lambda \eta_i - \eta_i) \ ,$$

so that we obtain the curves $\gamma_{i,n}$ verifying

8.19 $$\gamma_{i,n} \equiv \gamma_{i,n-1} \quad \text{mod. } \theta^n C \quad , \quad \Lambda\gamma_{i,n} \equiv \gamma_i \quad \text{mod. } \theta^{n+1} C \ ,$$

by solving the equations in $\bar{\eta}_i$:

8.20 $$\bar{\varepsilon}_i + f(\bar{\eta}_i) - \bar{\eta}_i = 0 \quad \text{in} \quad T \ ,$$

equivalent to a system of separable scalar equations in K . Therefore, we obtain for every i $(1 \leq i \leq \delta)$ a converging sequence $(\gamma_{i,n})_{n \in \underline{\mathbb{N}}}$ in C . The limits $\gamma_i = \lim_{n\to\infty} \gamma_{i,n}$ verify

8.21 $$\Lambda \cdot \gamma_i = \gamma_i \quad , \quad \bar{\gamma}_i = \bar{\gamma}_{i,o} \ .$$

As $(\gamma_i)_{1\leq i\leq \delta}$ is a basis of $C/\theta C$ over K , the curves $\theta^j \gamma_i$, $1 \leq i \leq \delta$, $0 \leq j < \mu$, are a V-basis of C , and <u>we have proved that the formal group</u> G <u>over</u> K <u>is isomorphic to</u> $(G_{\mu,\nu})^\delta$ <u>iff its module</u> $C = \mathfrak{C}_p(G)$ <u>verifies</u> (8.3). Its ring of endomorphisms is then a matrix ring $M_\delta(R)$ with entries in R (see 7.42).

CHAPTER VII

EXTENDING AND LIFTING SOME FORMAL GROUPS

1. <u>Extensions with additive kernels of formal groups of finite height</u>

1.1 Let G and G_1 be formal groups over a basic ring K . In any exact sequence of Cart(K)-modules

$$0 \to \mathfrak{C}(G_1) \to C_o \to \mathfrak{C}(G) \to 0 \ ,$$

the module C_o is reduced, because when we map into C_o a V-basis of $\mathfrak{C}(G_1)$ and complete it by lifting in C_o a V-basis of $\mathfrak{C}(G)$, we obtain a V-basis of C_o . Therefore C_o corresponds to a formal group, say $C_o = \mathfrak{C}(G_o)$; by definition, G_o is an extension of G by G_1 (or with kernel G_1). <u>The extension theory of formal groups is equivalent to that of their modules of curves.</u>

In the first two sections of this chapter, K will be a *perfect field of characteristic* p , and we keep the notations listed in (VI.1). We replace the general Cart(K)-modules by E-modules. An extension G_o of the formal group G by G_1 is defined by an exact sequence of E-modules,

1.2 $$0 \to C_1 \to C_o \overset{\alpha}{\to} C \to 0 ,$$

where $C_1 = \mathfrak{C}_p(G_1)$, $C_o = \mathfrak{C}_p(G_o)$, $C = \mathfrak{C}_p(G)$ (see VI.3.6).

As noted before (see VI.4.16), the corresponding sequence of $\tilde{E}$-modules,

1.3 $$0 \to \tilde{C}_1 \to \tilde{C}_o \overset{\tilde{\alpha}}{\to} \tilde{C} \to 0$$

is also exact.

1.4 Here we restrict our attention to a very special class of extensions. Namely we assume that G *is a formal group of finite height* h (see VI.7.9) and that *the kernel* G_1 *is an additive group* (see II.2.21).

1.5 *Under those assumptions, the exact sequence* (1.3) *splits, in the category of* $\tilde{E}$-*modules*.

Proof. It is a consequence of Fitting's lemma (see VI.5.7), but a direct argument is preferable. In the exact sequence of E-modules (1.2), the kernel of F acting on C_o is its submodule C_1 . For any $\gamma_o \in C_o$, $F\gamma_o \in C_o$ depends only on $\alpha(\gamma_o) \in C$ and even on $F\alpha(\gamma_o)$, for F is injective on C . Therefore there is a *unique* E-*isomorphism* β_o: FC → FC_o to insert in the commutative diagram

1.6 $$\begin{array}{ccc} C_o & \xrightarrow{F} & FC_o \\ {\scriptstyle\alpha}\downarrow & & \uparrow{\scriptstyle\beta_o} \\ C & \xrightarrow{F} & FC \end{array} ,$$

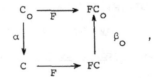

and $\alpha \circ \beta_o$ is the identity on FC.

But FC is a lattice in C (see VI.4.20), so that it has the same V-divided module $\tilde{C}$ as C , and β_o has a unique extension $\tilde{\beta}_o : \tilde{C} \to \tilde{C}_o$, which is an $\tilde{E}$-linear section of $\tilde{\alpha}$. The sequence (1.3) splits in a well defined way: any $\gamma_o \in \tilde{C}_o$ may be identified with a pair $(\gamma_1, \gamma) \in \tilde{C}_1 \times \tilde{C}$, where

1.7 $$\gamma = \tilde{\alpha}(\gamma_o) \quad , \quad \gamma_1 = \gamma_o - \tilde{\beta}_o(\gamma) \in \tilde{C}_1 \quad .$$

The exact sequence (1.2) does not necessarily split, because $\tilde{\beta}_o$ does not map necessarily $C \subset \tilde{C}$ into $C_o \subset \tilde{C}_o$.

The representatives in C_o of $\gamma \in C$ are identified with pairs $(\gamma_1, \gamma) \in \tilde{C}_1 \times C$, where γ_1 ranges over a class of $\tilde{C}_1$ modulo C_1. Therefore, the sequence (1.2) defines a map

1.8 $$\lambda : C \to \tilde{C}_1/C_1 \; ,$$

such that, for any $\gamma \in C$, $\gamma_1 \in \tilde{C}_1$,

1.9 $$(\gamma_1, \gamma) \in C_o \quad \text{iff} \quad \gamma_1 \in \lambda(\gamma) \; .$$

1.10 Proposition. Let $C = \mathbb{C}_p(G)$ and $C_1 = \mathbb{C}_p(G_1)$ be given, subject to the conditions of (1.4), and let us denote by $\operatorname{Ext}^1_E(C,C_1)$ the set of isomorphy classes of exact sequences (1.2); the isomorphisms have to induce identity on C and on C_1. Then there is a natural one-to-one correspondence between $\operatorname{Ext}^1_E(C,C_1)$ and the group $\operatorname{Hom}_E(C,\tilde{C}_1/C_1)$ of E-linear maps of C into $\tilde{C}_1/C_1$.

Proof. To any exact sequence (1.2) we associate the map $\lambda : C \to \tilde{C}_1/C_1$ defined by (1.9), and we check that λ is E-linear. Conversely, if the E-linear map λ is given, we define an E-submodule C_o of $\tilde{C}_1 \times C$ by formula (1.9), and it is immediately checked that the sequence (1.2) is exact.

1.11 Remark. The notation $\mathrm{Ext}_E^1(C,C_1)$ is that of Homological Algebra ([4], thm. XIV.1.1). The exact sequence

$$0 \to C_1 \to \tilde{C}_1 \to \tilde{C}_1/C_1 \to 0$$

gives birth to an exact sequence

1.12 $$\ldots \to \mathrm{Hom}_E(C,\tilde{C}_1) \to \mathrm{Hom}_E(C,\tilde{C}_1/C_1) \to \mathrm{Ext}_E^1(C,C_1) \to \mathrm{Ext}_E^1(C,\tilde{C}_1) \to \ldots$$

so that there is an alternative proof of proposition (1.10), by showing that

1.13 $$\mathrm{Hom}_E(C,\tilde{C}_1) = 0 \text{ , and}$$

1.14 $$\mathrm{Ext}_E^1(C,\tilde{C}_1) = 0 \text{ .}$$

If $f \in \mathrm{Hom}_E(C,\tilde{C}_1)$, then f has an unique $\tilde{E}$-linear extension $\tilde{f} : \tilde{C} \to \tilde{C}_1$. But F acts as an automorphism on $\tilde{C}$ and as 0 on $\tilde{C}_1$, which implies $\tilde{f} = 0$, and $f = 0$. As for (1.14), it means that any extension of C by $\tilde{C}_1$ splits. If we have an exact sequence of E-modules

$$0 \to \tilde{C}_1 \to \bar{C}_o \xrightarrow{\bar{\alpha}} C \to 0 \text{ ,}$$

we show as before that there is a unique E-linear section $\bar{\beta}_o : FC \to F\bar{C}_o$ of $\bar{\alpha}$ over FC, and we extend it to a section of α over the whole of C, because $\tilde{C}_1$ is V-divisible.

1.15 Now let us discuss what it means for a map $\lambda : C \to \tilde{C}_1/C_1$ to be E-linear. Commutation with the operator F means that

1.16 $$\lambda(FC) = 0 \text{ ,}$$

so that λ may be viewed as a map of C/FC into $\tilde{C}_1/C_1$. Then we have only to consider E-modules where F acts as 0, i.e. (E/FE)-modules. This ring will be denoted by $[[V]]K$, to remind that it is (identified with) the ring of twisted formal series $\Sigma_{n \in \underline{\underline{N}}} V^n a_n$, $a_n \in K$, with the commutation rule

1.17 $$aV = Va^p \quad , \quad a \in K . \qquad \text{(see VI.1.6, 1.7)} .$$

1.18 The ring $[[V]]K$ is a local ring, any of its ideals being generated by some power of V. The structure of the $[[V]]K$-modules of finite (Jordan-Hölder) length is easy to describe: they are direct sums of modules defined by a generator ε and a relation $V^n\varepsilon = 0$, for some $n \in \underline{\underline{P}}$. That is proved by a classical argument, already recalled when proving proposition (VI.5.20). The length is also the dimension over K.

1.19 Let us consider the ring $\{\{V\}\}K$ of meromorphic twisted power series, $\Sigma_{n_o \leq n < \infty} V^n a_n$, $n_o \in \underline{\underline{Z}}$, $a_n \in K$, which contains the subring $[[V]]K$. The additive factor group $\{\{V\}\}K/[[V]]K$, together with its natural structure of left-and-right $[[V]]K$-module, will be denoted by $\underline{\underline{U}}$. If M is a left (resp. right) module over $[[V]]K$, then the group of left (resp. right) linear maps of M into $\underline{\underline{U}}$, $\mathrm{Hom}(M,\underline{\underline{U}})$, has a natural structure of right (resp. left) module over $[[V]]K$: it will be called the dual of M, and denoted by M'.

1.20 From the structure of $[[V]]K$-modules of finite length (see 1.18), it is easy to deduce that such a module M is naturally isomorphic to its bidual $M'' = (M')'$.

1.21 An additive formal group G_1 is a direct sum of copies of $\underline{\underline{G}}_a$ (see II.2.13) and $C_1 = \mathfrak{C}_p(G_1)$ splits accordingly as a direct sum of copies of $C_{\underline{\underline{a}}} = \mathfrak{C}_p(\underline{\underline{G}}_a)$. There is a natural module-isomorphism between $C_{\underline{\underline{a}}}$ and $[[V]]K$, mapping $\gamma_{\underline{\underline{a}}}$ on the unit, so that $\tilde{C}_{\underline{\underline{a}}}/C_{\underline{\underline{a}}}$ may be identified with $\underline{\underline{U}}$ (see 1.19). Therefore we may reformulate proposition (1.10) as follows.

1.22 Proposition. The group $\mathrm{Ext}^1_E(C,C_{\underline{\underline{a}}})$ of extensions of a

<u>formal group of finite height</u> G <u>by the additive group</u> $\underline{\underline{G}}_a$ <u>is naturally isomorphic to the dual</u> $\mathrm{Hom}(C/FC,\underline{U}) = (C/FC)'$, <u>as defined in</u> (1.19).

2. The universal extension with additive kernel in characteristic p

2.1 Notations are as in section (1). Let d and d' be respectively the dimension and the codimension of G , so that $d + d' = h$ (the height of G), $d = \dim_K(C/V.C)$, $d' = \dim_K(C/F.C)$.

Qua W-module, $C = \mathfrak{C}_p(G)$ is free of rank h (see VI.7.9), and we may take a basis

2.2 $$\gamma_1,\ldots,\gamma_{d'}\ ,\ \gamma_{d'+1},\ldots,\gamma_h$$

of C over W , such that $\gamma_i \in F.C$ for $d' < i \leq h$. Then $F.C$ is the W-submodule of C with basis

2.3 $$p\gamma_1,\ldots,p\gamma_{d'}\ ,\ \gamma_{d'+1},\ldots,\gamma_h\ .$$

Indeed, all those elements lie in $F.C$ and they generate a submodule of codimension (over K) $d' = \dim_K(C/F.C)$.

2.4 <u>Lemma</u>. <u>Let</u>

$$0 \to C_1 \to C_o \xrightarrow{\alpha} C \to 0$$

<u>be an exact sequence of</u> E-<u>modules, where</u> $C = \mathfrak{C}_p(G)$ <u>and</u> $C_1 = \mathfrak{C}_p(G_1)$, G_1 <u>being an additive formal group. Then there is a</u> (W,F)-<u>linear section</u> $\beta : C \to C_o$ <u>of</u> α , <u>i.e. a map which is additive, commutes to</u> F <u>and operators of</u> W , <u>such that</u> $\alpha\circ\beta = \mathrm{Id}_C$.

Proof As C is free over W , we obtain a W-linear section just by choosing representatives $\beta(\gamma_i) \in C_o$ of the elements

(2.2) of a basis, and extending β to C by W-linearity. Then β is also F-linear iff its restriction to $F.C$ is the map β_o of diagram (1.6). So we have to put

2.5 $$\beta(\gamma_i) = \beta_o(\gamma_i) \quad \text{for} \quad d' < i \leq h ,$$

while we may take any representatives $\beta(\gamma_i)$ of γ_i for $1 \leq i \leq d'$, because

$$p\beta(\gamma_i) = F(V\beta(\gamma_i)) = \beta_o(p\gamma_i) .$$

Let

2.6 $$F.\gamma_i = \Sigma_{1\leq j\leq h}\, c_{i,j}\cdot\gamma_j \quad , \quad 1 \leq i \leq h \ , \quad c_{i,j} \in W ,$$

be the equations defining the action of F on C, given as a free W-module with the basis (2.2).

2.7 We define a formal group G^* of dimmension h by the V-basis $(\gamma_i^*)_{1\leq i\leq h}$ and the presentation

2.8 $$F.\gamma_i^* = \Sigma_{1\leq j\leq h}\, c_{i,j}\cdot\gamma_j^* \ , \ 1 \leq i \leq h \ , \quad \text{(see V.5.15)},$$

where the coefficients $c_{i,j} \in W$ are as in (2.6).

We took a basis of C to prove the existence of G^* by the structure theorem, but this group does not depend on the choice of the basis.

Indeed, putting $C^* = \mathfrak{C}_p(G^*)$, the equations (2.8) define C^* as the E-module freely generated by copies γ^* of the curves $\gamma \in C$, the relations in C being verified in C^* forasmuch as they imply only the (W,F)-module structure. Categorically, we apply the adjoint functor $C \to C^*$ of the forgetful functor "forget V".

2.9 In other words, C^* is an E-module, given with a (W,F)-linear map

2.10 $$\beta^* : C \to C^* \ , \quad (\gamma_i \mapsto \gamma_i^*) \ ,$$

with the following universal property: any (W,F)-linear map $\beta : C \to C_o$ of C into an E-module admits a unique factorization

2.11 $$\beta = u \circ \beta^* \ ,$$

where $u : C^* \to C_o$ is an E-linear map.

2.12 When we apply (2.11) to $\beta = Id_C$, we obtain the E-linear map

$$\alpha^* : C^* \to C \ ,$$

of which β^* is a (W,F)-linear section. We have an exact sequence

2.13 $$0 \to N \to C^* \to C \to 0 \ ,$$

corresponding to an exact sequence of formal groups.

2.14 Lemma. The kernel N in (2.13) corresponds to an additive formal group, i.e. $F.N = 0$.

Proof. Any element of the form

2.15 $$h_\gamma = \beta^*(V.\gamma) - V.\beta^*(\gamma) \quad , \quad \gamma \in C \ ,$$

verifies $F.h_\gamma = 0$, because β^* commutes to F and is additive. On the other hand, they generate the kernel N of α^*, because they express the "V-relations, $\gamma' = V.\gamma$ " of C, lifted in C^*.

Let us apply again (2.11) when β is a (W,F)-linear section of α, as in lemma (2.4). Then the E-linear map $u : C^* \to C_o$ verifies

2.16 $$\alpha \circ u = \alpha^* \ ,$$

because that is equivalent to $\alpha \circ u \circ \beta^* = \alpha^* \circ \beta^* = Id_C$. The submodule $u(N)$ of C_o is contained in the kernel of F, i.e. C_1, and we denote by $v : N \to C_1$ the restriction of u. To summarize, we have the following commutative diagram in the category of E-modules.

2.17
$$\begin{array}{ccccccccc}
0 & \to & C_1 & \to & C_o & \overset{\alpha}{\to} & C & \to & 0 \\
 & & \uparrow{\scriptstyle v} & & \uparrow{\scriptstyle u} & & \uparrow{\scriptstyle Id_C} & & \\
0 & \to & N & \to & C^* & \underset{\alpha^*}{\to} & C & \to & 0
\end{array}$$

2.18 Lemma. *Any* E-*linear map* $v : N \to C_1$ *defines an extension* C_o *of* C *by* C_1 , *together with the commutative diagram* (2.17), *up to unique isomorphism*.

Proof. The module C_o is obtained as a factor module of $C^* \times C_1$, the kernel being the set of elements $(\gamma,-v(\gamma)) \in C^* \times C_1$, $\gamma \in N$. One says that C_o is an amalgamated sum of C^* and C_1 over N , or that the left square in (2.17) is cocartesian.

2.19 Under the assumptions of (2.4), *the diagrams* (2.17) *are in one-to-one correspondence with the (non empty) set of* (W,F)-*linear sections* β *of* α . If we take two such sections β,β', then their difference $\beta'-\beta$ takes its values in C_1 , so that there is an E-homomorphism $w : C^* \to C_1$ verifying

2.20
$$\beta' - \beta = w \circ \beta^* .$$

The difference $v'-v$ of the E-homomorphism $v,v' : N \to C_1$ corresponding to β and β' is the *restriction to* N *of the* E-*linear map* $w : C^* \to C_1$. Conversely, v and w define v' and the corresponding section β' , $\beta' = u\circ\beta^* + w\circ\beta^*$. We have proved the following result.

2.21 Proposition. *There is a one-to-one correspondence between the set* $\mathrm{Ext}^1_E(C,C_1)$ *and the set* $\mathrm{Hom}_E(N,C_1)/\mathrm{Hom}_E(C^*,C_1)$ *of the*

E-*linear maps* $N \to C_1$ *modulo those which may be extended to* $C^* \to C_1$. *There is an exact sequence*

2.22 $$\mathrm{Hom}_E(C^*,C_1) \to \mathrm{Hom}_E(N,C_1) \to \mathrm{Ext}^1_E(C,C_1) \to 0 .$$

2.23 *Remark*. We did not prove that the group structure on $\mathrm{Ext}^1_E(C,C_1)$ defined by (2.22) is the same as in (1.22), but that follows from Homological Algebra [4]. The exact sequence (2.13) gives birth, generally, to an exact sequence

2.24 $$\ldots \to \mathrm{Hom}_E(C^*,C_1) \to \mathrm{Hom}_E(N,C_1) \to \mathrm{Ext}^1_E(C,C_1) \to \mathrm{Ext}^1_E(C^*,C_1) \to \mathrm{Ext}^1_E(N,C_1) \to \ldots$$

so that *the exactness of* (2.22) *is equivalent to the map*

2.25 $$\mathrm{Ext}^1_E(C^*,C_1) \to \mathrm{Ext}^1_E(N,C_1)$$

being injective.

Now, an element of $\mathrm{Ext}^1_E(C^*,C_1)$ is defined by an exact sequence

2.26 $$0 \to C_1 \to C_2 \xrightarrow{f} C^* \to 0 ,$$

which we combine with the exact sequence (2.13), turned vertically, to obtain the follow commutative diagram, with exact rows and columns:

2.27 $$\begin{array}{ccccccccc}
 & & & & & & 0 & & \\
 & & & & & & \downarrow & & \\
 & & 0 & & & & N & & \\
 & & \downarrow & & & & \downarrow & & \\
0 & \to & C_1 & \to & C_2 & \xrightarrow{f} & C^* & \to & 0 \\
 & & \downarrow & & \downarrow \mathrm{Id} & & \downarrow \alpha^* & & \\
0 & \to & C_1' & \to & C_2 & \to & C & \to & 0 \\
 & & \downarrow & & & & \downarrow & & \\
 & & N & & & & 0 & & \\
 & & \downarrow & & & & & & \\
 & & 0 & & & & & &
\end{array}$$

The module C_1' is the kernel of $\alpha^* \circ f$, and the left vertical sequence defines an element of $\mathrm{Ext}^1_E(N,C_1)$, which must be the one given by (2.25), because errors of sign spring mainly from connecting homomorphisms (and do not matter in questions of nullity). Therefore the kernel of the map (2.25) is characterized by the splitting of the left vertical sequence. Then $F.C_1^! = 0$, and by proposition (2.21) applied to the lower horizontal sequence, an E-linear map $C^* \to C_2$ may be introduced in (2.27), so that the sequence (2.26) splits.

2.28 Definition. Let G be a formal group of finite height. We call "universal extension of G with additive kernel" the formal group G^* defined in (2.7), together with the (W,F)-linear section $\beta^*: \mathbb{C}_p(G) \to \mathbb{C}_p(G^*)$.

2.29 Let us recall that, in diagram (2.17) with given upper sequence, the set of maps $u : C^* \to C_0$ is isomorphic to the set of (W,F)-linear sections β of α, by the relation $\beta = u \circ \beta^*$.

3. The reduction modulo p of p-typical groups

3.1 We shall now consider simultaneously two basic rings, K and $k = K/pK$. Strong conditions will be imposed later on K. For a while, we assume only that $pK \neq K$ (i.e. $k \neq 0$).

The natural map $\pi : K \to k$ is a "change of ring" (see II. 2.11, IV. 2.5), and its covariant action will be denoted generally by π_*.

We keep the notations listed in (VI.1), but the operators F and V will receive subscripts (K or k), to recall the basic ring under consideration.

3.2 As $\pi : K \to k$ is surjective, so are $\pi_*: E(K) \to E(k)$, $\pi_*: W(K) \to W(k)$ and, for any p-typical group G over K, the map $\pi_*: \mathfrak{C}_p(G) \to \mathfrak{C}_p(\pi_* G)$.

3.3 To lift over K a p-typical group Γ over k means to find a p-typical group G over K, together with an isomorphism $\pi_* G \to \Gamma$. That may be achieved by lifting in K the structural constants of Γ relative to some basic set of curves (see V. 5.15 and V.6.22).

The element

3.4 $$F_K V_K - V_K F_K = p - V_K F_K \in W(K) \subset E(K)$$

is mapped by π_* on $0 = F_k V_k - V_k F_k$. Moreover

3.5 $$\begin{cases} (p - V_K F_K) V_K = 0 , \\ F_K (p - V_K F_K) = 0 . \end{cases}$$

3.6 Let G be a p-typical group over K, and $\gamma \in \mathfrak{C}_p(G)$. By (3.5), the curve $\gamma' = (p - V_F F_K)\cdot\gamma$ depends only on $\mathfrak{T}\gamma \in \mathfrak{T}G$ and is additive, i.e. $F_K \cdot \gamma' = 0$; we have $\mathfrak{T}\gamma' = p\mathfrak{T}\gamma$.

3.7 Proposition. There is a morphism of p-typical groups over K,

$$\mathfrak{T}G^+ \to G ,$$

denoted by $x \mapsto \exp_G px$, mapping a curve $\Sigma_{n \in \underline{\underline{N}}} t^{p^n} a_n \in \mathfrak{C}_p(\mathfrak{T}G^+)$, $a_n \in \mathfrak{T}G$, on the curve $\Sigma_{n \in \underline{\underline{N}}} V_K^n (p - V_K F_K)\cdot\gamma_n$, where $\gamma_n \in \mathfrak{C}_p(G)$ is a representative of a_n ($a_n = \mathfrak{T}\gamma_n$) for any $n \in \underline{\underline{N}}$. The $E(K)$-homomorphism $\mathfrak{C}_p(\mathfrak{T}G^+) \to \mathfrak{C}_p(G)$ so defined will be denoted by ε_G.

Indeed, if $(\gamma_i)_{i \in I}$ is a V-basis in $\mathfrak{C}_p(G)$, then, putting $a_i = \mathfrak{T}\gamma_i$, we have a basis $(a_i)_{i \in I}$ of $\mathfrak{T}G$, and there is one morphism of p-typical groups, mapping the curve "$t \mapsto a_i t$" in $\mathfrak{T}G^+$ on $(p - V_K F_K)\cdot\gamma_i$, $i \in I$. This morphism does not depend on the choice of the basis.

3.8 From now on, we assume that K is p-torsion-free. So is the free K-module $\mathfrak{X}G$.

3.9 There is an exact sequence

$$0 \to \mathfrak{C}_p(\mathfrak{X}G^+) \xrightarrow[\varepsilon_G]{} \mathfrak{C}_p(G) \xrightarrow[\pi_*]{} \mathfrak{C}_p(\pi_* G) \to 0$$

Proof. Let $\gamma \in \mathfrak{C}_p(G)$, $\pi_* \gamma = 0$. Then $\mathfrak{X}\pi_* \gamma = 0$, which means that $\mathfrak{X}\gamma \in p\mathfrak{X}G$, and there are $\gamma', \gamma'' \in \mathfrak{C}_p(G)$ such that

$$\gamma = (p - V_K F_K) \cdot \gamma' + V_K \gamma'' \quad , \quad p\mathfrak{X}\gamma' = \mathfrak{X}\gamma \ .$$

As $\pi_*(V_K^n \cdot \gamma) = 0$ is equivalent to $\pi_* \gamma = 0$ for any $n \in \underline{\underline{N}}$, we prove by successive approximations that any $\gamma \in \mathfrak{C}_p(G)$ such that $\pi_* \gamma = 0$ may be written as

$$\gamma = \Sigma_{n \in \underline{\underline{N}}} V_K^n \varepsilon_G(a_n) \ ,$$

where $a_n \in \mathfrak{X}G$ is identified with the curve "$t \mapsto a_n t$" $\in \mathfrak{C}_p(\mathfrak{X}G^+)$. Moreover the vectors a_n are unique (for a given γ) iff K is p-torsion-free.

3.10 Definition. We call "logarithmic module" of a p-typical group G over a p-torsion-free ring K , and denote by $\underline{\underline{lm}}(G)$, the image of the map $\breve{D} : \mathfrak{C}_p(G) \to \mathfrak{X}G_p[[t]]$ (see V.7.3).

3.11 The map $\breve{D} : \mathfrak{C}_p(G) \to \underline{\underline{lm}}(G)$ is an isomorphism of $E(K)$-modules (see V.8.28), so that we may replace $\mathfrak{C}_p(G)$ by $\underline{\underline{lm}}(G)$ whenever it simplifies the formulas.

If $f = \Sigma_{n \in \underline{\underline{N}}} t^{p^n - 1} a_n \in \underline{\underline{lm}}(G)$, then

3.12
$$\begin{cases} V_K \cdot f = \Sigma_{n \in \underline{\underline{N}}} t^{p^{n+1} - 1} p a_n \ , \\ F_K \cdot f = \Sigma_{n \in \underline{\underline{N}}} t^{p^n - 1} a_{n+1} \ , \\ \xi \cdot f = \Sigma_{n \in \underline{\underline{N}}} t^{p^n - 1} w_n(\xi) a_n \ , \quad \text{when } \xi \in W(K) \end{cases}$$

(see VI.1.14) .

Therefore

3.13 $$(p-V_K F_K)\cdot f = pa_o \ .$$

3.14 Additive curves in G are characterized by the relation

$$\check{D}\gamma = \mathfrak{T}\gamma \quad \text{(qua formal series in } t) \ .$$

They are in one-to-one correspondence with their tangent vectors; the set of the latter contains $p\mathfrak{T}G$. It is equal to $\mathfrak{T}G$ iff G is an additive group (or equivalently a formal module); then

3.15 $$\Sigma_{n\in\underline{\underline{N}}}\, t^{p^n-1} a_n \in \underline{\underline{\mathrm{lm}}}(G) \quad \text{iff} \quad a_n \in p^n\mathfrak{T}G \ , \text{ for any } n \in \underline{\underline{N}}$$

By (3.9), the kernel of $\pi_*:\mathfrak{C}_p(G) \to \mathfrak{C}_p(\pi_*G)$ is characterized as follows. For $\gamma \in \mathfrak{C}_p(G)$, $\check{D}\gamma = \Sigma_{n\in\underline{\underline{N}}}\, t^{p^n-1} a_n$,

3.16 $$\pi_*\cdot\gamma = 0 \quad \text{iff} \quad a_n \in p^{n+1}\mathfrak{T}G \ , \text{ for any } n \in \underline{\underline{N}} \ .$$

3.17 Proposition. A p-typical group G over the p-torsion-free ring K is additive iff the formal group π_*G over $k = K/pK$ is additive.

Proof. The formal group π_*G is additive iff F_k vanishes on $\mathfrak{C}_p(\pi_*G)$, i.e. iff $\pi_*(F_k\cdot\gamma) = 0$ for any $\gamma \in \mathfrak{C}_p(G)$. Putting $a_n = \mathfrak{T}F_K^n\cdot\gamma \in \mathfrak{T}G$, we obtain from (3.12) and (3.16) the equivalent condition.

3.18 $$a_n \in p^n\mathfrak{T}G \ , \quad \text{for any } n \in \underline{\underline{N}} \ .$$

By (V.8.4) an (V.8.17), it means that $\log_G$ is defined over K , so that G is an additive group. An alternative proof may be obtained from (3.15).

4. On some ring homomorphisms $A \to W(A)$

In the present section we shall reproduce briefly some classical lemmas, before proving a criterion for Witt vectors (attributed by Cartier to Dwork and Dieudonné).

4.1 Let A be a p-torsion-free basic ring, so that any congruence

$$p^n x \equiv p^n y \quad \text{mod.} p^{m+n} \quad , \quad x,y \in A \ , \ m,n \in \underline{\underline{N}}$$

is equivalent to

$$x \equiv y \quad \text{mod. } p^m \ .$$

4.2 Lemma. Any congruence

$$x \equiv y \quad \text{mod. } p^m \quad , \quad \text{where} \quad x,y \in A, \ m \in \underline{\underline{P}}$$

implies

$$x^{p^n} \equiv y^{p^n} \quad \text{mod. } p^{m+n} \quad , \quad \text{for any} \quad n \in \underline{\underline{N}} \ .$$

That follows, for $n = 1$, from the binomial expansion, and the induction on n is immediate.

4.3 Lemma. Let x_i, y_i be given in A, for $0 \leq i \leq n$. Then the congruences

4.4 $$x_i \equiv y_i \quad \text{mod. } p \ , \qquad 0 \leq i \leq n \ ,$$

are equivalent to the congruences

4.5 $$\Sigma_{0 \leq i \leq m} \, p^i x_i^{p^{m-i}} \equiv \Sigma_{0 \leq i \leq m} \, p^i y_i^{p^{m-i}} \quad \text{mod. } p^{m+1}, \quad 0 \leq m \leq n \ .$$

Indeed, (4.4) implies

$$x_i^{p^{m-i}} \equiv y_i^{p^{m-i}} \quad \text{mod. } p^{m+1-i} \ ,$$

$$p^i x_i^{p^{m-i}} \equiv p^i y_i^{p^{m-i}} \quad \text{mod. } p^{m+1} \ ,$$

whence (4.5). Conversely, induction on n shows that (4.5) implies (4.4)

4.6 Criterion for Witt vectors. Let φ be an endomorphism of the p-torsion-free ring A, such that

4.7 $$\varphi(x) \equiv x^p \quad \text{mod. } p \ , \quad \text{for any} \quad x \in A \ ,$$

and let $(a_n)_{n \in \underline{\underline{N}}} \in A^{\underline{\underline{N}}}$. Then the system of equations

4.8 $$\Sigma_{o\leq i\leq n}\, p^i x_i^{p^{n-i}} = a_n \;,$$

<u>where</u> $n \in \underline{\underline{N}}$, <u>has a solution</u> $(x_n)_{n\in\underline{\underline{N}}} \in A^{\underline{\underline{N}}}$ <u>iff</u>

4.9 $$a_{n+1} \equiv \varphi(a_n) \mod. p^{n+1} \;, \quad \text{for any} \quad n \in \underline{\underline{N}} \,.$$

Proof. Let us assume that $x_n \in A$ has been computed for $0 \leq n \leq m$, so that (4.9) holds. Then the equation for x_{m+1} is

$$p^{m+1}\, x_{m+1} = a_{m+1} - \Sigma_{o\leq i\leq m}\, p^i x_i^{p^{m+1-i}} \;,$$

so that the criterion is true iff

4.10 $$\varphi(a_m) \equiv \Sigma_{o\leq i\leq m}\, p^i x_i^{p^{m+1-i}} \mod. p^{m+1} \,.$$

The relation (4.10) is proved by the same argument as lemma (4.3). Namely, $x_i^p \equiv \varphi(x_i) \mod. p$ implies

$$x_i^{p^{m+1-i}} \equiv \varphi(x_i)^{p^{m-i}} \mod. p^{m+1-i} \;,$$

$$p^i x_i^{p^{m+1-i}} \equiv p^i \varphi(x_i)^{p^{m-i}} \mod. p^{m+1} \;,$$

$$\Sigma_{o\leq i\leq m}\, p^i x_i^{p^{m+1-i}} \equiv \Sigma_{o\leq i\leq m}\, p^i \varphi(x_i)^{p^{m-i}} \mod. p^{m+1}$$

$$\equiv \varphi(a_m) \mod. p^{m+1} \;,$$

because φ is an endomorphism.

4.11 Let us take for A a polynomial ring over $\underline{\underline{Z}}$ in two sequences of indeterminates, $A = \underline{\underline{Z}}[x_o, x_1, \ldots ; y_o, y_1, \ldots]$ and define the endomorphism $\varphi : A \to A$ by $\varphi(x_i) = x_i^p$, $\varphi(y_i) = y_i^p$, $i \in \underline{\underline{N}}$. Then (4.7) is verified. Clearly, the sequences $(a_n)_{n\in\underline{\underline{N}}}$ in A verifying (4.9) form a <u>subring</u> of $A^{\underline{\underline{N}}}$ for its product ring structure. That proves the existence of polynomials with coefficients in $\underline{\underline{Z}}$ giving the sum and the product of two generic Witt vectors $x = (x_n)_{n\in\underline{\underline{N}}}$ and $y = (y_n)_{n\in\underline{\underline{N}}}$: this proof of the existence of the ring of (classical) Witt vectors is a simplification of Witt's original proof.

4.12 *Proposition. Let* A *and* φ *be as in* (4.6). *Then there is a ring homomorphism* $\Delta : A \to W(A)$, *defined by the relations*

4.13 $$w_n \circ \Delta = \varphi^n \quad \text{(the n-th iterated of } \varphi\text{), for any } n \in \mathbb{N} .$$

We have

4.14 $$(\Delta(x))^F = \Delta(\varphi(x)) \quad , \text{ for any } \quad x \in A .$$

Proof. The existence of the ring homomorphism Δ is an immediate consequence of (4.6), and (4.14) follows from (VI.1.9).

5. *A p-adic lemma*

5.1 *Lemma. Let* M *be an additive group, with two endomorphisms* V,F , *such that*

5.2 $$VF = FV = p\,\mathrm{Id}_M .$$

We assume that there is an integer $\nu \in \mathbb{P}$ *such that*

5.3 $$V^{\nu}M \subset p\,M .$$

Let T *be an additive group and* T' *a subgroup of* T . *We assume that* T *and* T/T' *are p-torsion-free, and that* T' *is complete for its p-adic topology , i.e.*

5.4 $$T' = \varprojlim (T'/p^nT') .$$

Let $\mu \in \mathbb{N}$ *be an integer and, for every* $n \in N$, *let* $f_n : M \to T$ *be an additive map such that*

5.5 $$f_{n+1} V \equiv p\, f_n \mod. p^{n+1+\mu}\, T \text{ , for any } n \in \mathbb{N}.$$

Then there is a unique additive map $g : M \to T$ *such that*

5.6 $$g\,F^n \equiv f_n \mod. p^{n+\mu}\, T' \text{ , for any } n \in \mathbb{N} .$$

Moreover, if M *and* T *are modules over some ring* A ,

<u>and if</u> φ <u>is an automorphism of</u> A <u>such that</u>

5.7 $$F(ax) = \varphi(a)F(x) \ , \ V(ax) = \varphi^{-1}(a)V(x) \ , \ f_n(ax) = \varphi^n(a)f_n(x) \ ,$$

<u>for any</u> $a \in A, x \in M, n \in \underline{\underline{N}}$, <u>then the map</u> g <u>of</u> (5.6) <u>is</u> A-<u>linear</u>.

5.8 Proof. We shall apply several times the following remark. Let $f, f' : M \to T$ be two additive maps such that, for some $i \in \underline{\underline{N}}$, $fV^i = f'V^i$ (resp. $fF^i = f'F^i$). Then $f = f'$. Indeed, the additive map $f - f'$ vanishes on V^iM (resp. F^iM), and a fortiori on p^iM for $p^i Id_M = V^iF^i = F^iV^i$. As T is p-torsion-free, that implies $f - f' = 0$.

By (5.5) and induction on n , we have

5.9 $$f_n V^n \equiv p^n f_o \quad \text{mod. } p^{n+\mu} T' \ , \text{ for any } n \in \underline{\underline{N}} \ .$$

Therefore, there are additive maps $g_n : M \to T$, such that

5.10 $$f_n V^n = p^n g_n = g_n F^n V^n \quad , \quad \text{for any } n \in \underline{\underline{N}} \ .$$

By (5.8), we have

5.11 $$f_n = g_n F^n \quad \text{for any } n \in \underline{\underline{N}} \ .$$

When substituting the values (5.11) of f_n in condition (5.5), we obtain

5.12 $$g_{n+1} \, F^{n+1} V \equiv p \, g_n \, F^n \quad \text{mod. } p^{n+1+\mu} \, T' \ , \text{ for any } n \in \underline{\underline{N}} \ ,$$

or equivalently

5.13 $$(g_{n+1} - g_n) F^n \equiv 0 \quad \text{mod. } p^{n+\mu} \, T' \ , \quad \text{for any } n \in \underline{\underline{N}} \ ,$$

because T and T/T' are p-torsion-free.

Therefore, there are additive maps $h_n : M \to T'$, defined by

5.14 $$(g_{n+1} - g_n) F^n = p^{n+\mu} \, h_n = p^{\mu} \, h_n V^n F^n \quad \text{for any } n \in \underline{\underline{N}} \ ,$$

so that, by (5.8),

5.15 $$g_{n+1} - g_n = p^{\mu} h_n V^n \quad , \quad \text{for any } n \in \underline{\underline{N}} .$$

It follows from (5.3) that $V^n M \subset p^m M$ when $n \geq \nu m$, so that, by (5.15), $g_{n+1} - g_n$ converges towards 0 for the p-adic topology on T'. As T' has been assumed complete (5.4), the sequence $(g_n)_{n \in \underline{\underline{N}}}$ has a limit $g : M \to T$. More precisely, we define the additive maps $k_n : M \to T'$ by

5.16 $$k_n = \Sigma_{i \in \underline{\underline{N}}} \; h_{n+i} \, V^i \; , \quad \text{for any } n \in \underline{\underline{N}} ,$$

and, by (5.15), we define the additive map $g : M \to T$ by

5.17 $$g - g_n = p^{\mu} k_n V^n \quad , \quad \text{for any } n \in \underline{\underline{N}} .$$

Therefore, we have

5.18 $$gF^n - g_n F^n = p^{n+\mu} \, k_n \quad , \quad \text{for any } n \in \underline{\underline{N}} ,$$

so that, by (5.11), we have the wanted relation (5.6).

To prove the unicity of g, let us assume that some additive map $g' : M \to T$ verifies also

5.19 $$g'F^n \equiv f_n \mod p^{n+\mu} \, T' \quad \text{for any } n \in \underline{\underline{N}} .$$

Then we have

5.20 $$(g-g')F^n = p^{n+\mu} \, l_n = p^{\mu} \, l_n V^n F^n \quad \text{for any } n \in \underline{\underline{N}} ,$$

where $l_n : M \to T'$ is an additive map. By (5.8), (5.20) is equivalent to

5.21 $$g - g' = p^{\mu} \, l_n V^n \quad \text{for any } n \in \underline{\underline{N}} ,$$

and it follows from (5.3) that $g - g'$ takes its values in $\cap_m p^m T'$, i.e. vanishes.

It follows from (5.7) that the maps $f_n V^n$ are A-linear for any $n \in \underline{\underline{N}}$. So are the maps g_n defined by (5.11), and so is their limit g for the p-adic topology.

6. Reduction from $W(k)$ to k (a perfect field of characteristic p)

6.1 We come back to the reduction modulo p of formal groups (see 3) to study a special case. Namely we assume that k is a perfect field of characteristic p and that $K = W(k)$, the ring of Witt vectors over k.

6.2 By (VI.3.13), we may identify k with the factor ring K/pK, and the natural map $\pi : K \to K/pK$ with the ring homomorphism $w_{o,k}: W(k) \to k$.

6.3 The automorphism $\xi \mapsto \xi^F$ of $W(k) = K$ verifies

$$\xi^F \equiv \xi^p \quad \text{mod. } p \qquad \text{(see VI.1.9)} ,$$

so that we may apply proposition (4.12) to the ring K. We have a ring homomorphism

6.4 $$\Delta : K \to W(K) ,$$

defined by the relations

6.5 $$(w_{n,K} \circ \Delta)(\xi) = \xi^{F^n} , \quad n \in \underline{\underline{N}} , \ \xi \in K .$$

When $n = 0$, (6.5) means that Δ is a ring-morphic section of $w_{o,K} = \pi_* : W(K) \to W(k) = K$. The ring $W(K)$ is a supplemented K-algebra in the sense of ([4], X.1), i.e.

6.6 $$W(K) = \Delta(K) \oplus \mathrm{Ker}_{W(K)} \pi_* .$$

6.7 Let us denote by G a finite-dimensional formal group over K, by $C = \mathbb{C}_p(G)$ the $E(K)$-module of p-typical curves, by Γ the formal group $\pi_* G$ over k and by M the $E(k)$-module $\mathbb{C}_p(\Gamma)$.

We have (see V.6.24), up to unique isomorphism,

6.8 $$M = E(k) \otimes_{E(K)} C ,$$

where the tensor product is defined by the ring homomorphism

$\pi_*: E(K) \to E(k)$. Then the map $\pi_* : C \to M$ is written $x \mapsto 1 \otimes x$, and may be said to be $E(K)$-linear, by considering an $E(k)$-module qua $E(K)$-module via $\pi_*: E(K) \to E(k)$.

The same procedure enables us to consider any $W(K)$-module, such as C , as a $W(k)$-module, i.e. a K-module, via Δ . As we have

6.9 $$\Delta(\xi^F) = (\Delta(\xi))^F \qquad \text{(see 4.14)} ,$$

we may extend the homomorphism $\Delta : W(k) \to W(K)$ to an homomorphism of the subring generated by $W(k)$ and F_k in $E(k)$ into the subring generated by $W(K)$ and F_K in $E(K)$ (see VI.1.9).

The tangent space $\mathfrak{T}G = C/V_K C$ is mapped by π_* onto the tangent space $\mathfrak{T}\Gamma = M/V_k M$; here the map π_* is just the reduction modulo p , i.e. its kernel is $p\mathfrak{T}G$. We have

6.10 $$\mathfrak{T}(\Delta(\xi).\gamma) = \xi\cdot\mathfrak{T}\gamma ,$$

for any $\xi \in K$, $\gamma \in C$, because $W(K)$ acts on $\mathfrak{T}G$ via $w_{o,K} : W(K) \to K$.

6.11 Let $(\gamma_{i,K})_{i\in I}$ be an indexed set in C , and $\gamma_{i,k} = \pi_*\gamma_{i,K}$, $i \in I$. Not only is $(\gamma_{i,k})_{i\in I}$ a V_k-basis of M when $(\gamma_{i,K})_{i\in I}$ is a V_K-basis of C , but also the converse is true. As $\mathfrak{T}G$ is <u>a free module of finite rank over the local ring</u> K <u>with radical</u> pK , Nakayama's lemma (or a direct argument) shows that a basis of $\mathfrak{T}G$ over K is just a set of elements, the images of which modulo $p\mathfrak{T}G$ are a basis of $\mathfrak{T}\Gamma = \mathfrak{T}G/p\mathfrak{T}G$.

6.12 <u>Definition</u>. We say that a map $f : M' \to C'$ of an $E(k)$-module M' into an $E(K)$-module C' is (W,F)-linear iff f is additive and

6.13 $$f(\xi\gamma) = \Delta(\xi)\cdot f(\gamma) , \quad f(F_k\cdot\gamma) = F_K.f(\gamma) ,$$

for any $\xi \in K$, $\gamma \in M'$.

6.14 Theorem. With the previous notations (see 6.7), let us assume that Γ is a formal group of finite height h over k . Then there is a unique (W,F)-linear section $\sigma : M \to C$ of $\pi_* : C \to M$.

Proof. Let $(\gamma_{i,k})_{1 \leq i \leq h}$ be a basis of M over K (see VI.7.4), and let $\gamma_{i,K} \in C$ be a representative of $\gamma_{i,k}$, $1 \leq i \leq h$ $(\pi_* \gamma_{i,K} = \gamma_{i,k})$. Then we define a section σ_o of $\pi_*: C \to M$ by putting

$$\sigma_o(\Sigma_{1 \leq i \leq h}\, \xi_i \cdot \gamma_{i,k}) = \Sigma_{1 \leq i \leq h}\, \Delta(\xi_i) \cdot \gamma_{i,K} \;, \quad \xi_i \in K \;.$$

We replace σ_o by $\breve{D} \circ \sigma_o : M \to \underline{\underline{\mathrm{lm}}}(G)$, and we write $(\breve{D} \circ \sigma_o)(\gamma)$ as a formal series

$$\Sigma_{n \in \underline{\underline{N}}}\, t^{p^n - 1}\, f_n(\gamma) \;, \qquad f_n : M \to \mathfrak{T}G \;.$$

The W-linearity of σ_o is expressed by the relations

$$f_n(\gamma+\gamma') = f_n(\gamma) + f_n(\gamma') \;, \quad n \in \underline{\underline{N}} \;, \quad \gamma, \gamma' \in M \;,$$

$$f_n(\xi\gamma) = \xi^{F^n} f_n(\gamma) \;, \quad n \in \underline{\underline{N}} \;, \ \xi \in K \;, \ \gamma \in M \;.$$

(see 3.12 and 6.5).

The map $\gamma \mapsto \sigma_o(V_k \cdot \gamma) - V_K \cdot \sigma_o(\gamma)$ takes its values in the kernel of $\pi_* : C \to M$, because $\pi_* V_K = V_k \pi_*$. By (3.12) and (3.16), that is expressed by the relations

$$f_{n+1}(V_k \cdot \gamma) \equiv p\, f_n(\gamma) \quad \text{mod. } p^{n+2}\, \mathfrak{T}G \;, \ n \in \underline{\underline{N}} \;, \ \gamma \in M \;.$$

All the assumptions of lemma (5.1) are verified, by putting $\mu = 1$, $V = V_k$, $F = F_k$, $T = T' = \mathfrak{T}G$, $\varphi(\xi) = \xi^F$. Therefore there is one, and only one, map $g : M \to \mathfrak{T}G$, which is K-linear and verifies

$$g(F_k^n \cdot \gamma) \equiv f_n(\gamma) \quad \text{mod. } p^{n+1}\, \mathfrak{T}G \;, \ n \in \underline{\underline{N}} \;, \ \gamma \in M \;.$$

By (3.16), there is a unique section $\sigma : M \to C$ such that

6.15 $\qquad (\overset{\lor}{D} \circ \sigma)(\gamma) = \Sigma_{n \in \underline{\underline{N}}} \, t^{p^{n-1}} \, g(F_k^n \cdot \gamma) \quad , \quad \gamma \in M ,$

and by (3.12) it is the unique (W,F)-linear section of $\pi_* : C \to M$.

Let, as in (2.6)

6.16 $\qquad F_k \cdot \gamma_{i,k} = \Sigma_{1 \le j \le h} \, c_{i,j} \cdot \gamma_{j,k} \quad , \quad 1 \le i \le h \, , \, c_{i,j} \in W(k) = K,$

be the equations defining the semi-linear operator F_k on the

free K-module M .

6.17 $\qquad$ <u>We define a formal group</u> G^* <u>of dimension</u> h <u>over</u> K <u>by</u>

<u>the</u> V_K-<u>basis</u> $(\gamma_{i,K}^*)_{1 \le i \le h}$ <u>and the presentation</u>

6.18 $\qquad F_K \cdot \gamma_{i,K}^* = \Sigma_{1 \le j \le h} \, \Delta(c_{i,j}) \cdot \gamma_{j,K}^* \quad , \quad 1 \le i \le h \qquad$ (see V.5.15) .

We put $C^* = \mathfrak{C}_p(G^*)$, and we define the (W,F)-linear map

6.19 $\qquad\qquad\qquad \tau : M \to C^* \qquad$ by

$\tau(\Sigma_{1 \le i \le h} \, \xi_i \cdot \gamma_{i,k}) = \Sigma_{1 \le i \le h} \, \Delta(\xi_i) \cdot \gamma_{i,K}^* \quad , \quad \xi_i \in K .$

By (V.6.22), the formal group $\pi_* G^*$ over k is the uni-

versal extension of Γ with additive kernel (see 2.28), which

we denote by Γ^* . So we have a commutative diagram

6.20

$$
\begin{array}{ccccccccc}
& & & & C^* & & & & \\
& & & \pi_* \downarrow & \nearrow & \tau & & & \\
0 & \longrightarrow & N_k & \longrightarrow & M^* & \underset{\beta^*}{\overset{\alpha^*}{\rightleftarrows}} & M & \longrightarrow & 0 \ ,
\end{array}
$$

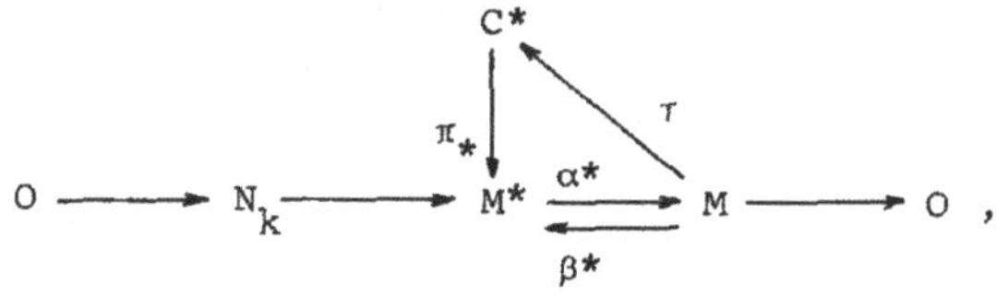

where $M^* = \mathfrak{C}_p(\Gamma^*)$, the kernel N_k being the $E(k)$-module of

p-typical curves in an additive group (see 2.14).

$\qquad$ <u>The diagram</u> (6.20) <u>depends only on</u> M (<u>i.e. on the formal</u>

<u>group</u> Γ), not on the choice of the V_k-basis $(\gamma_{i,k})_{1 \le i \le h}$ in

M . Indeed <u>the pair</u> (C^*, τ) <u>is defined, up to unique isomorphism,</u>

<u>by the following universal property.</u>

6.21 For any $E(K)$-module C' and any (W,F)-linear map $\sigma : M \to C'$ (see 6.12), there is a unique $E(K)$-linear map $\alpha : C^* \to C'$ such that $\sigma = \alpha \circ \tau$.

In particular, the universal map $\beta^* : M \to M^*$ is factorized as $\beta^* = \pi_* \circ \tau$ in diagram (6.20).

6.22 Proposition. The map $\iota : M \to \mathfrak{T}G^*$ defined by

6.23 $$\iota(\gamma) = \mathfrak{T}\tau(\gamma) \quad , \quad \gamma \in M ,$$

is a K-linear isomorphism, so that the map $\tau \circ \iota^{-1} : \mathfrak{T}G^* \to C^*$ is a section of the natural map $\mathfrak{T} : C^* \to C^*/V_K.C^* = \mathfrak{T}G^*$.

For any $\gamma \in M$, the formal series $(\check{D} \circ \tau)(\gamma)$ is

6.24 $$\Sigma_{n \in \underline{\underline{N}}} t^{p^n - 1} \iota(F_k^n \cdot \gamma) \in \underline{\underline{lm}}(G^*) .$$

Any curve $\gamma_K \in C^*$ has a unique expansion as

6.25 $$\gamma_K = \Sigma_{n \in \underline{\underline{N}}} V_K^n \cdot \tau(\gamma_n) \quad , \quad \gamma_n \in M ,$$

and verifies

6.26 $$\check{D} \gamma_K = \Sigma_{n \in \underline{\underline{N}}} t^{p^n - 1} \iota(a_n) \in \underline{\underline{lm}}(G^*) ,$$

where

6.27 $$a_n = \Sigma_{0 \leq i \leq n} p^i F_k^{n-i} \cdot \gamma_i \in M , \text{ for any } n \in \underline{\underline{N}} .$$

The logarithmic module of G^* is defined by

6.28 $$\Sigma_{n \in \underline{\underline{N}}} t^{p^n - 1} \iota(a_n) \in \underline{\underline{lm}}(G^*) \quad \text{iff} \quad F_k \cdot a_n \equiv a_{n+1} \mod . p^{n+1} M ,$$

6.29 Any additive curve $\gamma_K \in C^*$ (i.e. $F_K \cdot \gamma_K = 0$) may be written (in just one way) as $\gamma_K = \tau(V_k \cdot \gamma) - V_K(\tau \cdot \gamma)$, where $\gamma \in M$.

Proof. By (6.10), the additive map $\iota : M \to \mathfrak{T}G^*$ is K-linear and, by (6.19) it maps a basis of M over K onto a basis of $\mathfrak{T}G^*$ over K; therefore it is an isomorphism of K-modules.

Formula (6.24) is equivalent to the relations $\iota(\gamma) = \mathfrak{T}\tau(\gamma)$ and $\tau(F_k^n \cdot \gamma) = F_K^n \tau(\gamma)$, $\gamma \in M$, $n \in \underline{\underline{N}}$.

The existence and unicity of the expansion (6.25) for a given $\gamma_K \in C^*$ is equivalent to $\tau \circ \iota^{-1}$ being a section of $C^* \to C^*/V_K.C^*$.

The relations (6.26), (6.27) follow from (6.24), (6.25) by computation (see 3.12). They lead to the criterion (6.28) for $\underline{\underline{Im}}(G^*)$, because equations (6.27) may be rewritten as

6.30
$$p^{n+1}\gamma_{n+1} = a_{n+1} - F_k.a_n \quad , \quad n \in \underline{\underline{N}} .$$

Finally, $F_K.\gamma_K = 0$ iff $\breve{D}\gamma_K = \iota(a_o)$ (qua formal series) for some $a_o \in M$. By (6.28), a_o has to verify the condition $F_k.a_o \equiv 0$ mod. pM, i.e. $F_k.a_o = F_k V_k.\gamma$ or $a_o = V_k.\gamma$ for some $\gamma \in M$. Then (6.29) follows from (6.30).

7. Lifts from k over W(k)

7.1 Let K and k be as in (6.1), and Γ be a formal group over k. What we call a lift of Γ over K is, strictly speaking, a pair consisting of a formal group G over K and of an isomorphism $\pi_*G \to \Gamma$ of formal groups over k, or equivalently, of an isomorphism of $E(k)$-modules,

7.2
$$E(k) \otimes_{E(K)} \mathfrak{C}_p(G) \to \mathfrak{C}_p(\Gamma) .$$

To get rid of cumbersome notations, we shall write simply $\Gamma = \pi_*G$, but we must be careful about isomorphisms of lifts.

7.3 Two lifts G_1, G_2 of the same formal group Γ are said to be isomorphic iff there is an isomorphism of formal groups over K, $f : G_1 \to G_2$, inducing the identity on $\Gamma = \pi_*G_1 = \pi_*G_2$ (i.e. $\pi_*f = Id_\Gamma$). Then f is unique iff the natural group homomorphism

$$Aut_K(G_1) \to Aut_k(\Gamma)$$

is injective. We shall see presently that it is so when Γ has finite height, but it would not be so if Γ were, for instance, an additive group (see 3.17). Besides, if G_1, G_2 are non-isomorphic lifts of Γ, the (underlying) formal groups G_1, G_2 may still be isomorphic (i.e. there may be an isomorphism $G_1 \to G_2$ which does not induce the identity on Γ).

7.4 From now on, we assume again that Γ is a formal group of finite height h over k, and we keep all the notations of section (6). Starting from Γ, or equivalently from $M = \mathfrak{C}_p(\Gamma)$, we build up the formal group G^* over K, together with the (W,F)-linear map $\tau : M \to C^* = \mathfrak{C}_p(G^*)$, as in (6.17), (6.21).

7.5 Lemma. Let G be a lift of Γ over K and $C = \mathfrak{C}_p(G)$. Then there is a unique homomorphism of formal groups over K, $G^* \to G$, defined by the $E(K)$-linear map $f : C^* \to C$, such that $f \circ \tau : M \to C$ is a section of $\pi_* : C \to M$. Moreover, f is surjective.

Proof. By theorem (6.14), there is a unique (W,F)-linear section $\sigma : M \to C$ of $\pi_* : C \to M$. By (6.21), there is a unique $E(K)$-linear map $f : C^* \to C$ such that $\sigma = f \circ \tau$.

By (6.11), if $(\gamma_i)_{i \in I}$ is a V_k-basis of M, then $(\sigma(\gamma_i))_{i \in I}$ is a V_K-basis of C, so that $f(C^*)$ contains a V_K-basis of C, i.e. $f(C^*) = C$.

Now, by (3.9) and (7.5) we have, for any lift G of Γ, a commutative diagram with exact rows and columns:

7.6

$$\begin{array}{ccccccccc}
 & & & & 0 & & 0 & & \\
 & & & & \downarrow & & \downarrow & & \\
 & & & & \mathbb{C}_p(\mathfrak{T}G^{*+}) & \longrightarrow & \mathbb{C}_p(\mathfrak{T}G^{+}) & & \\
 & & & & \downarrow {\scriptstyle \varepsilon_{G^*}} & & \downarrow {\scriptstyle \varepsilon_G} & & \\
 & & & & C^* & \xrightarrow{f} & C & \longrightarrow & 0 \\
 & & & & \downarrow {\scriptstyle \pi_*} & & \downarrow {\scriptstyle \pi_*} & & \\
0 & \longrightarrow & N_k & \longrightarrow & M^* & \xrightarrow[\alpha^*]{} & M & \longrightarrow & 0 \\
 & & & & \downarrow & & \downarrow & & \\
 & & & & 0 & & 0 & &
\end{array}$$

As $f : C^* \to C$ is <u>surjective</u>, so is $\mathfrak{T}f : \mathfrak{T}G^* \to \mathfrak{T}G$ (because $\mathfrak{T}G^* = C^*/V_K.C^*$, $\mathfrak{T}G = C/V_K.C)$, and <u>the top horizontal map in diagram</u> (7.6) <u>is also surjective</u>.

If we denote by $N_K \subset C^*$ the kernel of f , then a classical "diagram chasing" shows that the restriction of π_* to N_K maps N_K <u>onto</u> N_k. So we have a commutative diagram with exact rows and columns

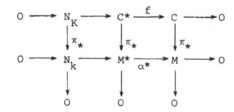

7.7

$$\begin{array}{ccccccccc}
0 & \longrightarrow & N_K & \longrightarrow & C^* & \xrightarrow{f} & C & \longrightarrow & 0 \\
 & & \downarrow {\scriptstyle \pi_*} & & \downarrow {\scriptstyle \pi_*} & & \downarrow {\scriptstyle \pi_*} & & \\
0 & \longrightarrow & N_k & \longrightarrow & M^* & \xrightarrow[\alpha^*]{} & M & \longrightarrow & 0 \\
 & & \downarrow & & \downarrow & & \downarrow & & \\
 & & 0 & & 0 & & 0 & &
\end{array}$$

The sequence

7.8 $$0 \to N_K/V_K.N_K \to C^*/V_K.C^* \to C/V_K.C \to 0$$

is exact, because C is V_K-torsion-free. Therefore, as C^*/V_KC^* is a free module over the principal ring $K = W(k)$, $N_K/V_K.N_K$ is also a free K-module. By (VI.1.12), the E(K)-module N_K is reduced, i.e. we may write $N_K = \mathbb{C}_p(A_K)$, where A_K is an <u>embedded subgroup of</u> G^* (see V.4.13), <u>and the lift</u> G <u>of</u> Γ <u>is naturally isomorphic to the factor formal group</u> G^*/A_K .

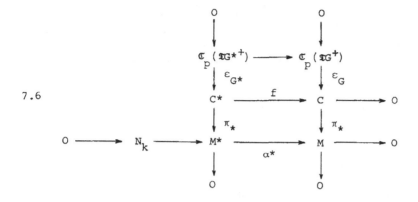

7.9 Let us denote by A_k the embedded subgroup of Γ^* such that $\mathfrak{C}_p(A_k) = N_k$. By (2.14), A_k is an additive group; more precisely, A_k is the *maximum additive (embedded) subgroup* of Γ^*, for N_k is the kernel of F_k on M^*.

The surjectivity of $\pi_*: N_K \to N_k$ in diagram (7.7) means that A_K is a lift of A_k. So, by (3.17), A_K *is an additive embedded subgroup of* G^*.

7.10 *Conversely, let* A_K *be an embedded subgroup of* G^*, *such that* $\pi_* A_K = A_k$, or equivalently $\pi_* N_K = N_k$, where $N_K = \mathfrak{C}_p(A_K) \subset C^*$. *Then the formal factor group* $G = G^*/A_K$, *defined by* $\mathfrak{C}_p(G) = C = C^*/N_K$ *is a lift of* Γ. That is quite clear when taking curvilinear coordinates in A_K and extending them to a system of curvilinear coordinates in G^*. Equivalently, we have a commutative diagram with exact rows and columns

7.11
$$\begin{array}{ccccccccc}
0 & \longrightarrow & N_K & \longrightarrow & C^* & \xrightarrow{\ f\ } & C & \longrightarrow & 0 \\
 & & \downarrow{\scriptstyle \pi_*} & & \downarrow{\scriptstyle \pi_*} & & & & \\
0 & \longrightarrow & N_k & \longrightarrow & M^* & \longrightarrow & M & \longrightarrow & 0 \\
 & & \downarrow & & \downarrow & & & & \\
 & & 0 & & 0 & & & &
\end{array}$$

where $f : C^* \to C$ is the natural map, and we introduce the surjective map $\pi_*: C \to M$, defining G as a lift of Γ (see 7.1).

7.12 The embedded subgroups A_K of G^* are in one-to-one correspondence with their respective tangent spaces $\mathfrak{T}A_K$, which are *direct summands* in the free K-module $\mathfrak{T}G^*$. By the reduction modulo p, $\mathfrak{T}G^* \to \mathfrak{T}\Gamma^* = \mathfrak{T}G^*/p\mathfrak{T}G^*$, the submodule $\mathfrak{T}A_K$ has to be mapped *onto the given submodule* $\mathfrak{T}A_k$ of $\mathfrak{T}\Gamma^*$. In other words, $\mathfrak{T}A_K + p\mathfrak{T}G^*$ *is a given submodule of* $\mathfrak{T}G^*$.

7.13 Lemma. $\mathfrak{T}A_K + p\mathfrak{T}G^* = \iota(V_k.M)$, where $\iota : M \to \mathfrak{T}G^*$ is the K-isomorphism of (6.23).

Proof. Any vector in $\mathfrak{T}A_K$ is tangent to an additive curve in G^* , because A_K is additive (see 7.10); so is any vector in $p\mathfrak{T}G^*$ (see 3.14). On the other hand, we know the set of all tangent vectors to additive curves in G^*: by (6.29), it is $\iota(V_kM)$. Therefore we have

$$\mathfrak{T}A_K + p\mathfrak{T}G^* \subset \iota(V_k.M)$$

To prove that we have equality in (7.13), it suffices to show that

7.14 $$\dim_K(\mathfrak{T}G^*/\mathfrak{T}A_K + p\mathfrak{T}G^*) = \dim_K(M/V_k.M) .$$

The left side is the codimension of $\mathfrak{T}A_k$ in $\mathfrak{T}\Gamma^*$, which is equal to the dimension of Γ , i.e. the right side of (7.14).

7.15 Conversely, if L is a direct summand in the K-module M , such that $L \subset V_k.M$, then $\iota(L)$ is the tangent space of an embedded additive subgroup A_K of G^*. Indeed, if $\{V_k\cdot\gamma_1,\ldots,V_k\cdot\gamma_r\}$ is a basis of L over K , then the additive curves $\{\tau(V_k\cdot\gamma_1) - V_k\tau(\gamma_1),\ldots,\tau(V_k\cdot\gamma_r) - V_K(\tau(\gamma_r)\}$ are a basic set in A_K .

7.16 Definition. The formal group G^* over K , together with the map $\tau : \mathfrak{C}_p(\Gamma) \to \mathfrak{C}_p(G^*)$, will be called the universal extended lift with additive kernel of Γ over K .

Our previous results may be summarized as follows.

7.17 Theorem. The isomorphy classes of the lifts over K of Γ (a formal group of finite height over k) are in one-to-one correspondence with the factor formal groups G^*/A_K , where A_K is a maximal embedded additive subgroup of G^* , and also with *the direct summands* L *of the* K-*module* $M = \mathfrak{C}_p(\Gamma)$ *such that* $L + pM = V_k.M$, *by the relation* $\mathfrak{t}A_K = \iota(L)$.

8. The universal extension with abelian kernel in characteristic 0

8.1 In this final section, notations are as in the two preceding ones, but all the data are kept fixed in the following diagram with exact rows and columns (see 7.11)

8.2
$$\begin{array}{ccccccccc} 0 & \longrightarrow & N_K & \longrightarrow & C^* & \xrightarrow{f} & C & \longrightarrow & 0 \\ & & \downarrow{\scriptstyle \pi_*} & & \downarrow{\scriptstyle \pi_*} & & \downarrow{\scriptstyle \pi_*} & & \\ 0 & \longrightarrow & N_k & \longrightarrow & M^* & \longrightarrow & M & \longrightarrow & 0 \\ & & \downarrow & & \downarrow & & \downarrow & & \\ & & 0 & & 0 & & 0 & & \end{array}$$

where $N_K = \mathfrak{C}_p(A_K)$, A_K being an additive formal group over K , of dimension d' (the codimension of Γ).

8.3 Let A be a finite-dimensional additive formal group (or formal module) over K , and $C_1 = \mathfrak{C}_p(A)$. Let

8.4
$$0 \longrightarrow C_1 \longrightarrow C_o \xrightarrow{f'} C \longrightarrow 0$$

be an exact sequence of $E(K)$-modules. Equivalently, $C_o = \mathfrak{C}_p(G_o)$ and G_o *is an extension of* G *with additive kernel* A (see 1.1).

8.5 *Theorem. With the above notation, there is a unique* $E(K)$-*linear map* $\varphi : C^* \to C_o$, *such that* $f = f' \circ \varphi$. *Equivalently, the group* $\mathrm{Ext}^1_{E(K)}(C, C_1)$ *is in natural isomorphism with the group* $\mathrm{Hom}_K(\mathfrak{t}A_K, \mathfrak{t}A)$ *of* K-*linear maps, and* $\mathrm{Hom}_{E(K)}(C^*, C_1) = 0$.

8.6 Proof. By (6.21) the existence and unicity of $\varphi : C^* \to C_o$ as in (8.5) is equivalent to the existence and unicity of a (W,F)-linear map, $\sigma' : M \to C_o$, such that $f' \circ \sigma' = \sigma$, the unique (W,F)-linear section of $\pi_* : C \to M$ (see 6.14).

As M is a free K-module, we may choose a map $\sigma_o' : M \to C_o$, which is additive, and verifies

8.7 $$f' \circ \sigma_o' = \sigma \quad ,$$

8.8 $$\sigma_o'(\xi \cdot \gamma) = \Delta(\xi)\sigma_o'(\gamma) \quad , \quad \text{for any} \quad \xi \in K \quad , \quad \gamma \in M \; .$$

Let us write $\check{D} \circ \sigma_o' : M \to \underline{\underline{lm}}(G_o)$ as

8.9 $$\gamma \mapsto \Sigma_{n \in \underline{\underline{N}}} \; t^{p^{n}-1} \; f_n(\gamma) \quad , \quad f_n : M \to \mathfrak{T}G_o \; .$$

We apply the p-adic lemma (5.1), putting $\mu = 0$, $T = \mathfrak{T}G_o$, $T' = \mathfrak{T}A \subset \mathfrak{T}G_o$, $V = V_k$, $F = F_k$. We have only to check that

8.10 $$f_{n+1}(V_k \cdot \gamma) - pf_n(\gamma) \in p^{n+1} \; \mathfrak{T}A \quad \text{for any} \quad \gamma \in M \; , \quad n \in \underline{\underline{N}} \; ,$$

then we shall obtain the unique K-linear map $g' : M \to \mathfrak{T}G_o$, such that

8.11 $$g'(F_k^n \cdot \gamma) - f_n(\gamma) \in p^n \; \mathfrak{T}A \quad , \quad \text{for any} \quad \gamma \in M \; , \quad n \in \underline{\underline{N}}$$

By (3.15) and (3.12), we shall have the unique (W,F)-linear map $\sigma' : M \to C_o$, defined by

8.12 $$\check{D} \circ \sigma' : \gamma \mapsto \Sigma_{n \in \underline{\underline{N}}} \; t^{p^{n}-1} \; g'(F_k^n \cdot \gamma) \; .$$

In order to prove (8.10), we remark that, for any $\gamma \in M$, the curve $\sigma(V_k \cdot \gamma) - V_k \cdot \sigma(\gamma) \in C$ is additive and must lie in the kernel of $\pi_* : C \to M$ because M is F_k-torsion-free. By (3.9), there is a map $\lambda : M \to \mathfrak{T}G$, such that

8.13 $$\sigma(V_k \cdot \gamma) - V_k \cdot \sigma(\gamma) - \varepsilon_G(\lambda(\gamma)) = 0 \quad , \quad \gamma \in M \; .$$

As $\mathfrak{T}f' : \mathfrak{T}G_o \to \mathfrak{T}G$ is surjective, we may choose a map $\lambda' : M \to \mathfrak{T}G_o$, such that $\mathfrak{T}f' \circ \mathfrak{T}\lambda' = \lambda$. Then, for any $\gamma \in M$ the curve

$$\sigma_o'(V_k \cdot \gamma) - V_k \cdot \sigma_o'(\gamma) - \varepsilon_{G_o}(\lambda'(\gamma))$$

lies in C_1, the kernel of f'. By applying the operator $\check{D}$, we find a series in $\underline{\underline{lm}}(G_1)$, and by (3.15) the relation (8.10) is verified.

The relation $\mathrm{Hom}_{E(K)}(C^*,C_1) = 0$ follows from the unicity of $\varphi : C^* \to C_o$ subject to $f'\circ\varphi = f$. The argument of (2.18) shows the isomorphism of $\mathrm{Ext}^1_{E(K)}(C,C_1)$ with $\mathrm{Hom}_{E(K)}(N_K,C_1)$. Finally, as $N_K = \mathbb{C}_p(A_K)$, $C_1 = \mathbb{C}_p(A)$ with additive formal groups (or formal modules) A_K, A over K , $\mathrm{Hom}_{E(K)}(N_K,C_1)$ is naturally isomorphic with $\mathrm{Hom}_K(\mathfrak{X}A_K,\mathfrak{X}A)$.

Quoted references

[1] BARSOTTI I. Metodi analitici per varietà abeliane in caratteristica positiva. Cap. 1,2 (Ann.Sc.Norm.Sup. Pisa Sc. Fis. Mat. ser. III, 18 (1964) p. 1-25).

[2] BLANCHARD A. Les corps non commutatifs (Paris, P.U.F., 1972).

[3] BOURBAKI N. Groupes et algèbres de Lie, Chapitre II (Paris, Hermann).

[4] CARTAN H. and EILENBERG S. Homological algebra (Princeton University Press, 1956).

[5] CARTIER P. Relèvements des groupes formels commutatifs (Sém. Bourbaki 1968/69, exposé 359, Lecture notes in Mathematics, n° 179).

[6] CONNEL I.G. Abelian formal groups (Proc. Am. Mat. Soc. 17 (1966) p. 958-959).

[7] DIEUDONNE J. Groupes de Lie et hyperalgèbres de Lie sur un corps de caractéristique $p > 0$. (Math.Ann. 134 (1957) p. 114-133).

[8] HAZEWINKEL M. Constructing formal groups, I,II,III,IV (Reports 7119 + appendix, 7201, 7207, 7322, Econometric Institute, Erasmus University, Rotterdam).

[9] LAZARD M. Sur les théorèmes fondamentaux des groupes formels commutatifs (Indag. Math. 35 (1973) p. 281-300 et 36 (1974 p. 122-124).

[10] Lois de groupes et analyseurs (Ann.Sci. Ec. Norm. Sup. Paris 72 (1955) p. 299-400).

[11] Leçons de calcul différentiel et intégral (à paraitre).

[12] LOOMIS L. H. An Introduction to Abstract Harmonic Analysis (Van Nostrand 1953).

[13] LUBIN J. and TATE J. Formal complex multiplication in local fields (Ann. of Math. 81 (1965) p. 380-387).

[14] SERRE J.-P. Commutativité des groupes formels de dimension 1 (Bull. Sci. Math. 91 (1967) p. 113-115).

[15] WEIL A. Basic number theory (Springer 1967).

[16] WITT E. Zyklische Körper und Algebren der Charakteristik p vom Grade p^n (J.f.r.u.a. Math. 176 (1937) p. 126-140).

Index

Artin-Hasse IV.9.19.

Basic.-ring I.1.1; - set of curves I.10.18.
Bud II.4.1.

Change of ring I.11.1.
Coboundary II.5.1.
Codimension of a formal group of finite height VII.7.9.
Composition operator I.10.10.
Coordinates I.4.1.
Curve I.6.2 ; canonical - III.3.21.
Curvilinear II.7.2.

Difference in degree q of two morphisms I.9.1.
Dimension of a formal variety I.6.12.

Embedded subgroup of a formal (or S-typical) group V.4.13.
Extension of formal groups VII.1.1.

Finite $\tilde{E}$-module VI.4.20.
Fitting's lemma VI.5.8.
Formal. - variety I.4.1; - module I.5.3; - group II.2.2.
Free uniform module V.I.3.

Generators of an uniform module V.1.2.
Group. - in a category II.1.1; formal - are commutative, unless otherwise stated II.2.1; - law, formal - II.2.2; additive - II.2.21.

Height of a formal group VII.7.9.
Hessian I.9.12.

Isoclinal. - automorphism VI.6.39; - formal group VI.7.9.
Isogenic VI.4.19.
Isogeny VI.4.21.

Jet I.2.4.

Lattice VI.4.20.
Law. group - II.2.2.
Length of an automorphism VI.6.12.
Lie algebra II.9.3.
Lift theorem II.9.3.
Local case IV.8.12.
Logarithmic module VII.3.10.

Model I.3.1.

Nilalgebra I.1.2.

Obstruction. homomorphism - II.5.2; bud - II.5.5.
Operator III.5.2.
Order. - topology I.2.6; - of a morphism I.5.12; - of an operator III.5.15; - function in an uniform module IV.5.5; - relative to a lattice VI.4.23; spectral - VI.6.17.

Presentation of an S-typical group V.5.3.
p-typical VI.1.3.

Reduced. - module III.7.15; - tensor product V.6.17; - derivative V.7.2.

Semi-linear endomorphism VI.5.4.
Simple. isogenically - formal group VI.7.18.
Slope IV.3.17.
S-torsion-free V.8.1.
Structural constants V.5.9.
S-typical IV.7.1; - group IV.7.4; - multiplicative group VI.9.1.

Tangent. - vector, - space I.6.3; - map I.6.4.
Topology. order - I.2.6; simple . I.2.7.
Type V.1.6.
Twisted formal series V.5.14.

Uniform module III.7.4.
Unipotent VI.5.19.
Universal. - group law V.10.29; - extension with additive kernel in characteristic p VII.2.28; - lift with additive kernel VII.7.16.

V-basis III.11.2 and IV.5.13.
V-divided module. relative - VI.3.15; absolute - VI.4.15.
V-divided ring VI.4.11.
V-divisible VI.3.9.
V-torsion-free VI.1.13.

(W,F)-linear VII.2.4. and VII.6.12.
Word II.1.2.
Witt vector III.1.14.

Table of notations

Chapter I $\underline{\underline{\underline{nil}}}(K)$ 1.2.; $\underline{\underline{\underline{nil}}}(K,n)$, $\underline{\underline{\underline{nilp}}}(K)$ 1.3.

J_n 2.4.; $\mathfrak{M}(V,W)$ 2.5.

$D^{(I)}$, D^n 3.1.

L^+ 5.3.; $\mathfrak{P}_n(L,M)$ 5.8.; $f \equiv f'$ mod.deg. (q+1) 5.11.; ord(f) 5.12.

$\mathfrak{C}$ 6.2.; $\mathfrak{T}, \mathfrak{T}\gamma, \mathfrak{T}V$ 6.3.; $\mathfrak{T}f$ 6.4. $f(x,y) \equiv x+y$ mod.deg.2 6.10.;

[c] 6.11.

$\mathrm{dif}_q(f,f')$ 9.2.

$\mathrm{comp}(\varphi)$ 10.10.; V_n 10.12.

φ_*, $\varphi_* f$ 11.1.

Chapter II Hom(G,G') 1.11.

$\mathfrak{M}(V,G)$ 2.8.; $\underline{\underline{G}}_a$ 2.13.; $\underline{\underline{G}}_m$ 2.14.

$\log_G$ 3.5.; $\exp_G$ 3.6.

δ 5.1.; Δ 5.2.; Γ_n 5.6.

B_n 5.15.; C_n 5.17.

Chapter III $\gamma_{\underline{\underline{a}}}$, $\gamma_{\underline{\underline{m}}}$ 1.3.; W^+ 1.14.; $\hat{W}^+$ 1.16.

∂ 2.3.; $\underline{\underline{w}}$, w_n 2.8.

F_n 3.6.; γ_w 3.21.

Cart(K) 5.10.

Chapter IV $M_{(I)}(K)$ 1.9.; κ 1.10.

S, $\underline{\underline{S}}$ 2.1.; $Cart_S$ 2.2.

κ' 3.11.; Sl S 3.17.

W_S 4.1. op 4.2.; $u_{k,n}$ 4.10.

$\tilde{M}$, $\kappa_{T,S}$ 6.6.

$\mathfrak{G}_S$ 9.1.; $\underline{\underline{Z}}_{T-S}$ 9.7.

Chapter V E 1.1.; tp 1.7.

$\geqslant_{sl}$ 2.5.; $>_{sl}$ 2.6.

$+_\mu$ 4.4.

$Cart_S(K)_+$ 5.11.

$\varphi_{G,A}$ 6.8.; $\underline{\otimes}$ 6.17.; $\psi_{G,A}$ 6.19.

$\breve{D}$ 7.2.; $M_S[[t]]$ 7.8.

K_S, M_S 8.1.; $\log_G$, $\int \check{D}\gamma$ 8.5.

Chapter VI E, E_+, W, F, V 1.3.; ξ^F 1.9.

x^F 2.11.

$div_c C'$ 3.16.

$\tilde{E}$ 4.11.; $\tilde{C}$ 4.15.; $ord_c(\gamma)$ 4.24.; lg(C,C') 4.27.

Λ 5.4.

$ord_c(\Lambda)$ 6.3.; $\mathfrak{T}_c$ 6.7.; lg(Λ) 6.12.; ord_{sp} 6.17.

ord_p 7.19., E_1 7.26.; $G_{\mu,\nu}$ 7.27.

Chapter VII $Ext^1_E(C,C_1)$ 1.10.; [[V]]K, {{V}}K, $\underline{\underline{U}}$ 1.16.

G^* 2.9.; β^* 2.10.

π_* 3.1.; $\exp_G px$, ε_G 3.7.; $\underline{\underline{\mathrm{lm}}}$, $\mathfrak{L}G_p[[t]]$ 3.10.

Δ 4.12.

G, C, Γ, M 6.7.; G*, Γ* 6.17.; ι 6.22.